# STIMULATING TECHNOLOGICAL PROGRESS

A STATEMENT BY THE RESEARCH AND
POLICY COMMITTEE OF THE
COMMITTEE FOR ECONOMIC DEVELOPMENT
JANUARY 1980

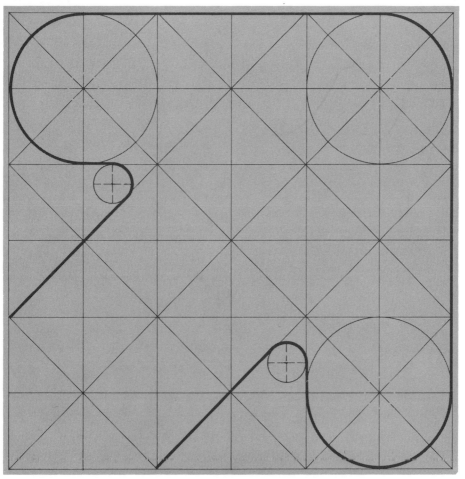

**Library of Congress Cataloging in Publication Data**

Committee for Economic Development. Research and Policy Committee.
   Stimulating technological progress.

   1. Technological innovations—United States.
2. United States—Economic policy—1971-
I. Title.
HC110. T4C65    1980    338.973    79-25164
ISBN 0-87186-770-2 lib. bdg.
ISBN 0-87186-070-8 pbk.

First printing: January 1980
Paperbound: $5.00
Library binding: $6.50
Printed in the United States of America by The Heffernan Press Inc.
Design: Stead, Young & Rowe, Inc.

COMMITTEE FOR ECONOMIC DEVELOPMENT
477 Madison Avenue, New York, N.Y. 10022
1700 K Street, N.W., Washington, D.C. 20006

# CONTENTS

| | |
|---|---|
| RESPONSIBILITY FOR CED STATEMENTS ON NATIONAL POLICY | vi |
| **PURPOSE OF THIS STATEMENT** | ix |
| **1. INTRODUCTION AND SUMMARY OF RECOMMENDATIONS** | 1 |
| Encouraging Technological Progress | 2 |
| Summary of Major Policy Recommendations | 4 |
| Strategy for Technology in the U.S. Economy | 9 |
| **2. TECHNOLOGY AND THE U.S. ECONOMY** | 11 |
| Importance of Technological Progress | 11 |
| Sources of Technological Progress | 13 |
| Phases of Technological Progress | 13 |
| Risks and Rewards | 14 |
| Financing Technological Progress | 16 |
| Technological Progress: A Coordinated Approach | 17 |
| **3. THE ECONOMY AND FUTURE TECHNOLOGICAL PROGRESS** | 19 |
| Economic Climate and Disincentives to Innovation | 19 |
| Policies to Revitalize Technological Progress | 20 |
| **4. TAX POLICIES FOR INVESTMENT AND INNOVATION** | 29 |
| Impact of General Tax Measures on Investment and Innovation | 30 |
| Impact of Special Tax Measures on Research and Development and Innovation | 34 |
| Toward Better Selective Tax Incentives | 36 |
|     Special U.S. Tax Provisions Affecting Invention and Innovation | 36 |
| Reducing the Cost of Invention and Innovation | 38 |
|     Foreign versus U.S. Tax Treatment of Invention and Innovation | 38 |
| Additional Tax Changes for Policy Consideration | 40 |
| **5. GOVERNMENT CONSTRAINTS ON INNOVATION** | 43 |
| Effects of Excessive Government Regulation | 43 |
| Regulatory Reforms to Stimulate Innovation | 46 |
| International Technology Transfer | 47 |
| National Security: A Special Case | 49 |
| **6. PATENTS AND THE INNOVATIVE PROCESS** | 51 |
| Role of Patents | 52 |
| Changes In Patent Policy to Enhance the Innovative Climate | 53 |
| Additional Recommendations on Patent Policy | 56 |

| | |
|---|---|
| **7. DIRECT GOVERNMENT SUPPORT OF RESEARCH AND DEVELOPMENT** | 58 |
| Trends in Government R & D Sponsorship | 58 |
| Guiding Principles for Government Support of Research and Development | 61 |
| Basic Research and the Universities | 63 |
| Federal Support for Large-Scale Applied R & D Programs | 64 |
| Federal R & D Performance | 66 |
| Conclusions | 67 |
| **MEMORANDA OF COMMENT, RESERVATION, OR DISSENT** | 68 |
| **APPENDIX A** <br> SUPPORTING ANALYSES FOR ADDITIONAL TAX CHANGES FOR POLICY CONSIDERATION | 76 |
| **APPENDIX B** <br> SUPPORTING ANALYSES FOR ADDITIONAL SUGGESTED CHANGES IN PATENT POLICY | 81 |
| OBJECTIVES OF THE COMMITTEE FOR ECONOMIC DEVELOPMENT | 86 |

# STIMULATING TECHNOLOGICAL PROGRESS

# RESPONSIBILITY FOR CED STATEMENTS ON NATIONAL POLICY

The Committee for Economic Development is an independent research and educational organization of two hundred business executives and educators. CED is nonprofit, nonpartisan, and nonpolitical. Its purpose is to propose policies that will help to bring about steady economic growth at high employment and reasonably stable prices, increase productivity and living standards, provide greater and more equal opportunity for every citizen, and improve the quality of life for all. A more complete description of CED is to be found on page 86.

All CED policy recommendations must have the approval of the Research and Policy Committee, trustees whose names are listed on page vii. This Committee is directed under the bylaws to "initiate studies into the principles of business policy and of public policy which will foster the full contribution by industry and commerce to the attainment and maintenance" of the objectives stated above. The bylaws emphasize that "all research is to be thoroughly objective in character, and the approach in each instance is to be from the standpoint of the general welfare and not from that of any special political or economic group." The Committee is aided by a Research Advisory Board of leading social scientists and by a small permanent professional staff.

The Research and Policy Committee is not attempting to pass judgment on any pending specific legislative proposals; its purpose is to urge careful consideration of the objectives set forth in this statement and of the best means of accomplishing those objectives.

Each statement is preceded by extensive discussions, meetings, and exchanges of memoranda. The research is undertaken by a subcommittee, assisted by advisors chosen for their competence in the field under study. The members and advisors of the subcommittee that prepared this statement are listed on page viii.

The full Research and Policy Committee participates in the drafting of findings and recommendations. Likewise, the trustees on the drafting subcommittee vote to approve or disapprove a policy statement, and they share with the Research and Policy Committee the privilege of submitting individual comments for publication, as noted on pages vii and viii and on the appropriate page of the text of the statement.

*Except for the members of the Research and Policy Committee and the responsible subcommittee, the recommendations presented herein are not necessarily endorsed by other trustees or by the advisors, contributors, staff members, or others associated with CED.*

# RESEARCH AND POLICY COMMITTEE

*Chairman:*
FRANKLIN A. LINDSAY

*Vice Chairmen:*
JOHN L. BURNS/*Education and Social and Urban Development*
E. B. FITZGERALD/*International Economy*
HOWARD C. PETERSEN/*National Economy*
WAYNE E. THOMPSON/*Improvement of Management in Government*

A. ROBERT ABBOUD, Chairman
The First National Bank of Chicago
¹ROY L. ASH, Chairman
AM International, Inc.
JOSEPH W. BARR, Corporate Director
Washington, D.C.
HARRY HOOD BASSETT, Chairman
Southeast Banking Corporation
JACK F. BENNETT
Senior Vice President
Exxon Corporation
CHARLES P. BOWEN, JR.
Honorary Chairman
Booz, Allen & Hamilton, Inc.
JOHN L. BURNS, President
John L. Burns and Company
FLETCHER L. BYROM, Chairman
Koppers Company, Inc.
ROBERT J. CARLSON
Group Vice President
United Technologies Corporation
RAFAEL CARRION, JR.
Chairman and President
Banco Popular de Puerto Rico
WILLIAM S. CASHEL, JR.
Vice Chairman
American Telephone
& Telegraph Company
JOHN B. CAVE, Senior Vice President,
Finance and Administration
Schering-Plough Corporation
EMILIO G. COLLADO, President
Adela Investment Co., S. A.
ROBERT C. COSGROVE, Chairman
Green Giant Company
RICHARD M. CYERT, President
Carnegie-Mellon University
W. D. DANCE, Vice Chairman
General Electric Company
JOHN H. DANIELS, Chairman
National City Bancorporation
W. D. EBERLE, Special Partner
Robert A. Weaver, Jr. and Associates
WILLIAM S. EDGERLY
Chairman and President
State Street Bank and Trust Company
FRANCIS E. FERGUSON, President
Northwestern Mutual Life
Insurance Company

JOHN H. FILER, Chairman
Aetna Life and Casualty Company
WILLIAM S. FISHMAN, Chairman
ARA Services, Inc.
E. B. FITZGERALD
Milwaukee, Wisconsin
DAVID L. FRANCIS, Chairman
Princess Coals, Inc.
¹JOHN D. GRAY, Chairman
Omark Industries, Inc.
H. J. HEINZ II, Chairman
H. J. Heinz Company
¹RODERICK M. HILLS
Latham, Watkins and Hills
ROBERT C. HOLLAND
President
Committee for Economic Development
EDWARD R. KANE, President
E. I. du Pont de Nemours & Company
PHILIP M. KLUTZNICK
Klutznick Investments
¹RALPH LAZARUS, Chairman
Federated Department Stores, Inc.
FRANKLIN A. LINDSAY, Chairman
Itek Corporation
¹J. PAUL LYET, Chairman
Sperry Corporation
G. BARRON MALLORY
Jacobs Persinger & Parker
WILLIAM F. MAY, Chairman
American Can Company
THOMAS B. McCABE
Chairman, Finance Committee
Scott Paper Company
GEORGE C. McGHEE
Corporate Director
and former U.S. Ambassador
Washington, D.C.
JAMES W. McKEE, JR., Chairman
CPC International Inc.
E. L. McNEELY, Chairman
The Wickes Corporation
J. W. McSWINEY, Chairman
The Mead Corporation
RUBEN F. METTLER, Chairman
TRW, Inc.
¹ROBERT R. NATHAN, Chairman
Robert R. Nathan Associates, Inc.

VICTOR H. PALMIERI, Chairman
Victor Palmieri and Company
Incorporated
HOWARD C. PETERSEN
Philadelphia, Pennsylvania
C. WREDE PETERSMEYER
Bronxville, New York
R. STEWART RAUCH, JR.
Chairman, Executive Committee
General Accident Group of
Insurance Companies
JAMES Q. RIORDAN
Executive Vice President
Mobil Oil Corporation
WILLIAM M. ROTH
San Francisco, California
¹HENRY B. SCHACHT, Chairman
Cummins Engine Company, Inc.
ROBERT B. SEMPLE, Chairman
BASF Wyandotte Corporation
RICHARD R. SHINN, President
Metropolitan Life Insurance Company
ROCCO C. SICILIANO, Chairman
Ticor
ROGER B. SMITH
Executive Vice President
General Motors Corporation
CHARLES B. STAUFFACHER, President
Field Enterprises, Inc.
WILLIAM C. STOLK
Weston, Connecticut
WILLIS A. STRAUSS, Chairman
Northern Natural Gas Company
WALTER N. THAYER, President
Whitney Communications Corporation
WAYNE E. THOMPSON
Senior Vice President
Dayton Hudson Corporation
SIDNEY J. WEINBERG, JR., Partner
Goldman, Sachs & Co.
GEORGE L. WILCOX, Director-Officer
Westinghouse Electric Corporation
¹FRAZAR B. WILDE, Chairman Emeritus
Connecticut General
Life Insurance Company
RICHARD D. WOOD, Chairman and
President
Eli Lilly and Company

¹Voted to approve the policy statement but submitted memoranda of comment, reservation, or dissent or wished to be associated with memoranda of others. See pages 68-75.

NOTE/*A complete list of CED trustees and honorary trustees appears at the back of the book. Company or institutional associations are included for identification only; the organizations do not share in the responsibility borne by the individuals.*

## SUBCOMMITTEE ON TECHNOLOGY POLICY IN THE UNITED STATES

*Chairman*
THOMAS A. VANDERSLICE
General Telephone & Electronics
 Corporation

DAVID BERETTA
Uniroyal, Inc.
JOHN C. BIERWIRTH
Grumman Corporation
DEREK BOK
Harvard University
ALFRED BRITTAIN III
Bankers Trust Company
FLETCHER L. BYROM
Koppers Company, Inc.
ROBERT D. CAMPBELL
Newsweek, Inc.
WILLIAM S. CASHEL, Jr.
American Telephone and
 Telegraph Company
RICHARD M. CYERT
Carnegie-Mellon University
JOHN DIEBOLD
The Diebold Group, Inc.

HARRY J. GRAY
United Technologies Corporation
FREDERICK G. JAICKS
Inland Steel Company
EDWARD R. KANE
E.I. du Pont de Nemours & Co.
CHARLES N. KIMBALL
Midwest Research Institute
JEAN MAYER
Tufts University
C. PETER McCOLOUGH
Xerox Corporation
THOMAS O. PAINE
Northrop Corporation
[1]D.C. SEARLE
G.D. Searle & Company
ROBERT B. SEMPLE
BASF Wyandotte Corporation
[1]MARK SHEPHERD, JR.
Texas Instruments, Inc.
DAVID S. TAPPAN, JR.
Fluor Corporation

HOWARD S. TURNER
Turner Construction Company
GEORGE L. WILCOX
Westinghouse Electric Corporation
J. KELLEY WILLIAMS
First Mississippi Corporation
RICHARD D. WOOD
Eli Lilly and Company

*Nontrustee Members
NORTON BELKNAP
Senior Vice President
Exxon International Corporation
JERRIER A. HADDAD
Vice President – Technical
 Personnel Department
IBM Corporation
GERALD D. LAUBACH
President
Pfizer, Inc.
GEORGE M. LOW
President
Rensselaer Polytechnic Institute

[1]Voted to approve the policy statement but submitted memoranda of comment, reservation, or dissent or wished to be associated with memoranda of others.
*Nontrustee members take part in all discussions on the statements but do not vote on it.

## ADVISORS TO THE SUBCOMMITTEE

RICHARD L. GARWIN
IBM Fellow and Science Advisor to
 the Director of Research
John Fitzgerald Kennedy School of
 Government
Harvard University
TAIT S. GOLDSCHMID
Senior Economic Analyst
Exxon Corporation
WALTER HAHN
Senior Specialist, Science and
 Technology and Futures Research
Congressional Research Service
Library of Congress
J. HERBERT HOLLOMON
Director, Center for Policy Alternatives
Massachusetts Institute of Technology

RONALD M. KONKEL
Office of Management and Budget
 (on leave)
WESLEY A. KUHRT
United Technologies Corporation
MAX MAGNER
Staff Consultant/Technical
 Government Liaison
E.I. du Pont de Nemours and Company
HARRY MANBECK
General Patent Counsel
General Electric Company
BOYD McKELVAIN
Staff Associate – Technology
 Policy Development
General Electric Company

RUDOLPH G. PENNER
Resident Scholar
American Enterprise Institute for
 Public Policy Research
ROLF PIEKARZ
Senior Staff Associate
Division of Policy Research
 and Analysis
National Science Foundation
ROGER SEYMOUR
Program Director, Commercial
 Relations
IBM Corporation

## PROJECT DIRECTOR
EDWIN S. MILLS
Chairman, Department of Economics
Princeton University

## CED STAFF ADVISORS
KENNETH McLENNAN
Vice President and Director of
 Industrial Studies
FRANK W. SCHIFF
Vice President and Chief Economist

## PROJECT EDITOR
CLAUDIA P. FEUREY
Director of Information
Committee for Economic Development

## PROJECT STAFF
LORRAINE MACKEY
Administrative Assistant

# PURPOSE OF THIS STATEMENT

During this century, technological change has been the primary source of economic growth in this country. New products and processes and better and more efficient methods of production have made the promise of a better life a reality for generations of Americans.

Over the past decade, there have been serious signs that the United States is losing its technological lead. For example:

- Capital investment as a percentage of output is now lower in the United States than in other major industrialized countries.
- A large share of industry's resources has been diverted away from investment in innovative activities toward compliance with government regulations.
- Other industrial nations, especially Japan and West Germany, are rapidly diminishing the U.S. technological lead, and some countries have surpassed the United States in a number of manufacturing areas.

Inflation, low productivity, trade deficits, and job loss are all linked to inadequate technological growth and innovation. Without technological progress, industry creates few jobs, factories and equipment become obsolete, productivity stagnates, and inflation becomes more difficult to control. If industry cannot produce goods and services more efficiently, it becomes less competitive in international markets and cannot compensate for rising costs except by raising prices.

**INVESTMENT IS THE KEY**

Early in the course of this study, it became clear that a declining rate of investment in plant and equipment was a major factor in the discouraging decline in technological growth. It also became clear that certain public policies actively discourage investment in new productive facilities, cause business to be defensive rather than creative, and reduce incentives for research and development.

Under the chairmanship of Thomas A. Vanderslice, the CED Subcommittee on Technology Policy examined a wide spectrum of policies which influence the entire technological process. We concluded that what is needed is a combined strategy for tax changes, regulatory reform, patent policy modifications, and increased support for basic research at universities to spur business investment in innovative and inventive products and processes. While all of these areas are important, we place the highest priority on specific changes in tax law to create greater incentives for investment. Our recommendations in these areas are discussed in detail in the following chapters.

Our recommendations pay particular attention to strengthening incentives for individual investors, small entrepreneurs and businesses, and to sources of high-risk capital to support new ventures. The report also stresses that greater technological progress is essential if the United States hopes to hold its own in increasingly competitive international markets.

In preparing our recommendations on regulation, the subcommittee drew extensively on the proposals for regulatory reform contained in CED's recent statement, *Redefining Government's Role in the Market System.*

## OUR ULTIMATE RESOURCE

While rising inflation and slowing productivity have brought technology policy into public debate, it is important to take a longer range look at the broader benefits that science and technology can bring. In a world with a growing population, rising social and economic expectations, and declining renewable resources, science and technology are our ultimate resources.

Firms that have invested heavily in developing technology and carrying it forward into commercial products have been shown to have about twice the productivity rate, three times the growth rate, nine times the employment growth, and one-sixth the price increases as firms with relatively low investment in these activities. This experience underscores how important technological progress can be to *all* American industry and to the achievement of some of this nation's most important social and economic goals.

It is vitally important that policy makers realize that the future quality of all our lives is directly tied to the scientific and technological progress we make today. However, while praising its benefits, we also recognize its potential hazards. Great industrial and technical strides have sometimes caused pollution, new health problems, and social disruption. But if progress has brought problems — and it has — then the answer is not less knowledge, but more. Technology, properly directed, can help social policy deal with these problems, and it can increase benefit-cost ratios in the process.

With free scientific pursuit and a growing technological base, there is every reason to be optimistic about new discoveries and new innovations helping us solve such problems as energy, disease, and pollution. I agree with scientist Philip Handler who wrote, "I see no alternative but to address vigorously the principal questions of science itself and to use our ever-widening understanding, our increasingly sophisticated technology with grace, charity, and wisdom. We are not omnipotent, but neither are we the foils of powerful forces over which we lack control. . . . Science is the principal tool that our civilization has developed to mitigate the condition of man."

**SPECIAL CONTRIBUTIONS**

The subcommittee that prepared this statement included a number of trustees and advisors with deep and extensive knowledge of the complex technical issues we examined. A list of all subcommittee members and advisors appears on page viii.

We are especially indebted to the chairman of the subcommittee, Thomas A. Vanderslice, president of General Telephone & Electronics Corporation, for his skill, expertise, and commitment to excellence. Additional recognition is also due to project director Edwin S. Mills for his rigorous approach to this subject and to staff counselor Kenneth McLennan and advisor Boyd McKelvain for their special contributions to the success of the project.

We are also grateful to The Andrew W. Mellon Foundation, the Lilly Endowment, Inc., Eli Lilly and Company, the Northrop Corporation, and Pfizer, Inc. for the generous support they provided which made this project possible.

Franklin A. Lindsay, *Chairman*
*Research and Policy Committee*

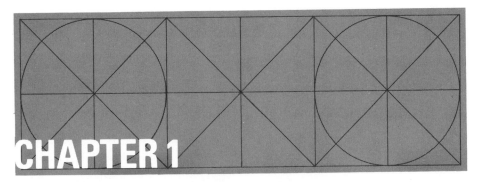

# CHAPTER 1
# INTRODUCTION AND SUMMARY OF RECOMMENDATIONS

A rising standard of living has long been a goal to which most Americans could reasonably aspire. This expectation of a better life has been supported and made possible by a strong and productive U.S. economy. Pacing that economic growth and productivity improvement was the unparalleled accomplishment of most sectors of the American economy in technological innovation.

That accomplishment enabled American industry to introduce a constant stream of new and improved products and services, to create new jobs, and to improve the productivity of both labor and capital. It has also enabled the United States to compete effectively in international trade and for a long time contributed to a favorable balance of payments. For many years, this continually increasing productivity enabled the country to cope with inflation, the bane of most industrial nations since the end of World War II. Economists have shown that more than half of the increase in American productivity has been due to scientific and engineering advances, industrial improvements, and management and labor know-how. Today, there is strong evidence that *technological progress is perhaps the most important source of future economic vitality and social progress for the United States.*[1]

---

[1] The term *technological progress* encompasses a number of interrelated activities, including basic research, applied research, development of commercial feasibility, innovation, and the diffusion of innovation. For further discussion of these activities, see Chapter 2.

From the end of World War II to 1964, U.S. labor productivity in the private business economy increased at an annual rate of over 3 percent; but over the past several years, the annual growth rate has declined to *less than 1 percent*. The overall *level* of productivity in the U.S. economy is higher than the level in all other industrial nations. But in the manufacturing sector of the economy, which has always been in the forefront in using technological innovation to raise productivity, this country's international competitors are rapidly catching up. Between 1960 and 1978, the average annual rate of productivity improvement for manufacturing in the United States was 2.8 percent compared with 8.2 percent in Japan and 5.4 percent in Germany. The rate for the United States was also lower than the rate for the United Kingdom (2.9 percent).

Recent productivity improvements have been insufficient to offset the cost of wage increases, and the resulting rising labor costs have contributed to inflationary pressures. Growth in real incomes has slackened. *This country cannot reasonably hope to control inflation, raise real incomes, and improve the quality of living unless the unfavorable trend in productivity is reversed.*

In some instances, technological change may impose undesirable short-term adjustment costs and undesirable side effects that must be taken into account. But more often than not, the best way to deal with adverse side effects is through additional innovations that eliminate or ameliorate them. On balance, the overall effect of technological change will be new employment opportunities and greater economic and social progress.

## ENCOURAGING TECHNOLOGICAL PROGRESS

The slowing of productivity improvement during the past few years parallels the discouraging decline in the rate of investment in plant and equipment. From 1968 to 1978, the annual average increase in plant and equipment investment was only 2.6 percent in real terms, compared with 6.2 percent in the previous decade. The data in Figure 1 show that the United States has lagged behind most other major industrial countries in capital investment in manufacturing as a proportion of the gross domestic product (GDP)[2] of manufacturing companies. The U.S. rate of investment as

---

[2] *Gross Domestic* Product (GDP) rather than Gross *National* Product (GNP) is used for comparing output per employed person among countries. GDP represents the final output of goods and services produced by a country's economy by residents and nonresidents. In contrast, GNP includes incomes (wages, interest, dividends, etc.) from abroad accruing to residents and excludes incomes earned in the domestic economy accruing to persons abroad. Since the comparative performance of economies is the main focus of this policy statement, GDP is the appropriate statistic to use.

a proportion of GDP has averaged about one-half the rate for France and Germany and about one-third the rate for Japan.

This decline is serious because investment in capital equipment not only improves productivity and stimulates employment opportunities but also encourages investment in inventive and innovative activity. Although total U.S. research and development (R & D) expenditures are much larger than any other nation's, a large portion of this country's effort is directed toward defense and space programs. These federally supported programs were sharply curtailed during the late 1960s and early 1970s, and private R & D efforts have not risen sufficiently to offset the government reductions from their peak in the mid-1960s. In 1976, industry-funded research and development was 1.05 percent of the gross national product (GNP) in the United States, 1.18 percent in West Germany, and 1.25 percent in Japan. Because of the rapidly increasing sophistication of modern techniques and equipment used in research, the cost of maintaining the existing level of R & D effort is rising at a rapid rate. Over the past several years, the focus of industrial research and development has shifted toward more conservative, shorter-term targets.

### FIGURE 1

**Average Annual Capital Investment and Rate of Growth in Output in Manufacturing**

| Country | Capital Investment as Proportion of Output* (Average Annual Percent, 1960–1976) | Growth in Output per Employee Hour** (Average Annual Percent Increase, 1960–1978) |
|---|---|---|
| United States | 9.1 | 2.8 |
| United Kingdom | 13.5 | 2.9 |
| France | 19.2[a] | 5.5 |
| Canada | 14.7 | 4.0 |
| Germany | 15.9[b] | 5.4 |
| Japan | 28.8[c] | 8.2 |

*For comparative purposes output is measured at current factor cost.
**All employed persons for U.S. and Canada; all employees for other countries.
[a] For total economy.
[b] 1960-1976.
[c] 1960-1974.
SOURCE: U.S. Department of Labor, Bureau of Labor Statistics.

Why has U.S. investment in the sources of technological progress lagged so badly? *This Committee believes the main cause is that some current public policies actually discourage new capital investment. They have also caused business to shift some of its R & D efforts from longer-term to more defensive, short-term goals.*\*

During the 1970s the relationship between the risk and the potential reward for longer-term, high-risk investments has changed. Certain policies have increased the risks of investment and have reduced potential rewards. Increased risk and uncertainty have resulted from the expansion and administration of government regulatory activities and uncertain energy and economic policies; reduced potential rewards have resulted from a combination of current tax policies and the failure to control inflation. Consequently, funds are being channeled away from more uncertain innovative opportunities and toward more certain consumption-oriented and hedging types of investments instead.

**We feel that unless there is speedy correction of public policies that inhibit the development and application of new products and processes within the market system, the nation will find it increasingly difficult to solve its other economic problems and achieve its social objectives.**

## SUMMARY OF MAJOR POLICY RECOMMENDATIONS

On the basis of this Committee's analyses of the issues, we recommend a series of changes in current public policies.

### TAX POLICY CHANGES

There is strong evidence that since 1973, a low rate of capital formation has contributed substantially to lower productivity growth.[3] Part of the problem is attributable to the combination of existing general tax policies and inflation. Federal tax revenue as a percentage of GNP was 18.5 percent in 1976; it will be about 20 percent in 1979 and is expected to be about 21 percent in 1981. The rising effective tax rate on business has inhibited capital investment.

Innovation through new capital investment is strongly influenced by the level of retained business earnings and the expected real rate of return on investment. Unfortunately, since the early 1970s, high rates of inflation have reduced the incentive to invest in new plant and equipment because allowable depreciation of existing plant and equipment is based on historical cost, which in a period of rapid inflation is much lower than the replace-

---

[3] J.R. Norsworthy and M.J. Harper, *The Role of Capital Formation in the Recent Slowdown in Productivity Growth*, U.S. Department of Labor, Bureau of Labor Statistics Working Papers, January 1979.

---

\*See memorandum by ROBERT R. NATHAN, page 68.

ment cost. In addition, the book values of inventories reflect a capital gain that is not an actual economic gain. As a result, taxable profits are higher than true economic profits. An important implication of this adverse effect of inflation is that the effective corporate tax rate in an inflationary period is considerably higher than the reported average corporate tax rate.

Figure 2 compares the trends in reported and effective corporate tax rates since 1960. It is quite clear that the high inflation rates during the pe-

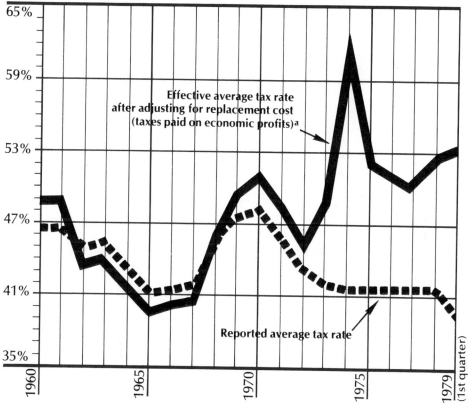

**FIGURE 2**

**Impact of Inflation on Effective Corporate Tax Rate, 1960 to 1979**

[a] In order to take into account two inflationary effects, the following adjustments are made: (1) Before-tax profits are reduced by the increase in depreciation that would result if plant and equipment were valued at replacement cost rather than original cost (capital consumption allowance), and (2) the capital gain from the increase in inventory prices due to inflation is eliminated.

SOURCE: U.S. Department of Commerce, Bureau of Economic Analysis.

riod from 1973 to 1975 caused a significant rise in the effective tax rate and had a detrimental impact on investment in new plant and equipment. Since 1977, the effective tax rate has again started to rise. If inflation continues into the 1980s, especially if it is accompanied by a recession, the recent increase in capital investment will surely fade once again; and consequently, the rate of technological innovation and productivity improvement will be adversely affected.*

The rate of return on investment for U.S. nonfinancial corporations is equally discouraging for future technological innovation. The real economic rate of return is the return on total capital investment (equity and

**FIGURE 3**

**Trend in Return on Total Capital Investment (Original versus Replacement Cost Basis) for Nonfinancial Corporations, 1960 to 1978**

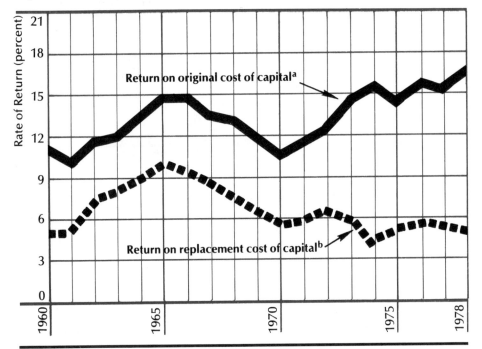

[a] Aftertax profits plus interest divided by original cost of plant and equipment.
[b] Aftertax profits plus interest minus capital consumption allowance and inventory gains divided by replacement cost of capital.
SOURCE: U.S. Department of Commerce, Bureau of Economic Analysis.

*See memorandum by MARK SHEPHERD, JR., page 68.

long-term debt) when investment is measured on the basis of the replacement cost of capital. Figure 3 shows that since 1960, the real rate of return based on replacement cost has been lower than that based on the historical cost of capital. Between 1960 and 1965, the difference in these rates was about 6 percentage points with the real rate of return being about 10 percent in 1965. However, since 1970, the difference between the real economic rate of return and the return measured by historical cost of capital has widened substantially. The difference has been about 11 percentage points, and the economic rate of return on total investment has declined to about 5 percent.

The willingness of business to invest in innovation depends on the expectation of a reasonable real return on the investment. Consequently, the recent trend and the low level of the return are not encouraging for future technological progress. **We believe it is essential that the government quickly adopt tax policies which increase the return on investment and stimulate future economic progress.**

This Committee believes that correcting disincentives in the tax system should be a major component of a comprehensive strategy for stimulating technological change.* **We conclude that there is an urgent need for more rapid capital recovery allowance as the number-one priority measure to stimulate investment in new plant and equipment.**\*\* We support the intent of current legislative proposals to shorten the capital recovery period for tax purposes, rather than to hold to the traditional concept that requires depreciation to be spread over the entire useful life of an asset. In order to overcome some of the investment disincentives caused by inflation, investors should be permitted capital write-offs that approximate the rising costs of replacing their plant and equipment.\*\*\*

Certain special tax measures are also needed to make investment in research and development more attractive. At present, tax provisions allow a deduction for depreciation of buildings, equipment, and other depreciable property used in R & D facilities that is equal to the deduction permitted on nonbusiness assets. **These depreciation provisions could be amended to allow flexible depreciation of all such R & D assets in order to take into account the inherent uncertainty of the usefulness of R & D assets.** Under such a system, the taxpayer would have the option of depreciating these assets fully in the first year of their life or adopting any other method desired while retaining the benefits of the allowable investment tax credit.

Other changes in the structure and level of tax rates may also be desirable, but such changes will have to take into account the effect on other policies concerned with inflation and overall spending.\*\*\*\*

---

*See memorandum by RODERICK M. HILLS, page 69.

---

\*\*See memorandum by RALPH LAZARUS, page 69.

---

\*\*\*See memorandum by ROBERT R. NATHAN, page 70.

---

\*\*\*\*See memorandum by MARK SHEPHERD, JR., page 70.

## REDUCING REGULATORY CONSTRAINTS ON INNOVATION

Regulatory policy changes are needed to make investment in innovation more rewarding. Although it recognizes that some government regulation of economic activity is necessary, the Committee has found that in many cases substantial compliance costs are significantly reducing the resources available for the technological innovation that increases productivity. Zero-risk goals, requirements for the use of best-available technology, and the frequent changes in production standards that are sometimes required have all contributed to escalating compliance costs. In addition, regulatory delays and court challenges to the legality of regulatory policies have reduced the rate of innovation. The resulting uncertainty about the acceptability of advanced technological applications has reduced the probability of any future gain from investments in innovation.

Although many of the goals of regulation are laudable, excessively detailed rules and specifications are too often used to carry out regulatory policies. Realistic performance standards should be set, but businesses seeking the most cost-effective method of production should be allowed the freedom to meet those standards in their own way.

CED's recent policy statement *Redefining Government's Role in the Market System*[4] calls for improvements in the congressional oversight process to curtail unjustified regulations and to reshape those that are justified. It urges that regulations be evaluated for their impact on the economy. On the basis of that analysis, the Committee places a high priority on the need to consider the impact of regulatory policies on technological innovation as a means of achieving the nation's economic objectives. We therefore support the efforts of the Administration and Congress to achieve regulatory reform.

We especially encourage the development of guidelines for determining whether new or existing regulations are needed and of periodic after-the-fact reviews of both the effectiveness and the economic impact of all forms of regulation. A systematic process for accomplishing these purposes is set forth in *Redefining Government's Role in the Market System*.

## REFORMING PATENT POLICY

The patent system was designed to stimulate industrial innovation. The protection provided by patents encourages the investment of funds not only in research and development but also in facilities to commercialize the R & D output. We propose a number of changes in the system that should

---

[4] For a more extensive discussion of recommendations on regulatory policy, see Chapters 6 and 7 of *Redefining Government's Role in the Market System* (1979).

increase its effectiveness and strengthen its supportive role in the innovation process.

One key area of improvement deals with the *resolution of disputes* over issued patents. The cost and time currently required to resolve contested situations seriously detract from the prompt and effective functioning of the system. Three changes are suggested: First, arbitration should be endorsed by statute as an acceptable way of settling patent controversies. Second, a single court of appeals for patent cases should be established to provide nationwide uniformity in the patent law. Third, a statutory reexamination procedure should be instituted which, at the request of parties other than the patentee, enables the Patent and Trademark Office (PTO) to strike obviously invalid patents from the rolls.

The *timing of the patent grant* is crucial to its reliability in business planning. Here, two changes are important: First, to protect innovation adequately in fields subject to government regulation, a procedure should be established providing for an appropriate adjustment in the patent term when commercialization is held up because of regulatory delay. Second, to prevent extended controversies and long delays in the issuance of patents when two or more inventors claim the same improvement, the nation should change to a first-to-file system, whereby the first inventor to file a patent application would receive the patent. (A personal right of use could be preserved for anyone filing later who in fact invented first and took steps leading to commercialization.) Under such a system, the ownership of all patents would be determined promptly, and the public would benefit from early publication of the patent disclosure.

## DIRECT FEDERAL SUPPORT OF RESEARCH AND DEVELOPMENT

The very large program of federally supported research and development is extremely important to industrial innovation. The Committee favors moves toward increased government support for basic research, but we recommend that federal involvement in selection and management of technological development aimed at commercial application should be undertaken only under extremely limited circumstances. Specific criteria for such decision making are proposed in Chapter 7.

## STRATEGY FOR TECHNOLOGY IN THE U.S. ECONOMY

This Committee believes that an increased rate of technological progress in industry is essential to achieving the nation's economic and social objectives. But although technological innovation can contribute broadly to future economic strength, business confidence in the future of the economy is a prerequisite for an adequate level of private investment in the longer-term, high-risk ventures that lead to innovation and its diffusion throughout the economy.

High and uncertain rates of inflation have severely constrained the nation's real economic growth. One of the underlying causes of this country's persistent inflation is its low rate of productivity improvement. Productivity can be significantly increased through the more rapid application of advanced technology in industrial processes. In our view, stimulation of a higher rate of investment in the economy's productive base will create more rapid technological progress and at the same time have a lasting impact on productivity improvement and inflation control.

Our strategy has three principal elements:

● First, the level of investment in plant and equipment should be raised in order to increase the diffusion of new technology into industrial processes. This would provide the structural change necessary for permanent impact on productivity growth and thereby on the control of inflation. We recommend that this be accomplished immediately by removing certain existing tax disincentives (which are discussed in Chapter 4).

● Second, nonessential regulatory constraints on, and uncertainties inherent in, productive investments should be reduced.* This, in conjunction with the improved economic performance that will result from more rapid productivity growth, would create the essential climate in which investment in all phases of technological innovation would be increased as the natural response of the entrepreneurial process. (These steps are discussed in Chapter 5.)

● Third, appropriate tax, patent, and regulatory changes should be made to provide support to foster private research and development. In addition, adequate support of *basic* research should be a high-priority item in the federal budget.

---

*See memorandum by D. C. SEARLE, page 71.

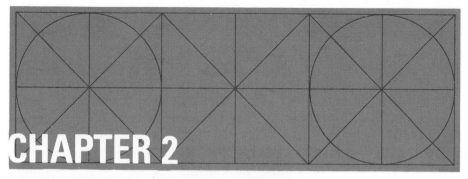

# CHAPTER 2
# TECHNOLOGY AND THE U.S. ECONOMY

Technological progress is among the most important and complex economic issues facing this country. Three decades ago, the United States was the world's undisputed technological leader, and until the last decade, the U.S. economy was characterized by a rapid rate of innovation. Gradually, however, other industrialized countries have improved their technological performances while U.S. technological progress has faltered. In the process, American citizens have lost potential increases in income, jobs, and enhancement of the quality of their lives.

## IMPORTANCE OF TECHNOLOGICAL PROGRESS

Technological progress enables the economy to produce new or improved commodities and services and more of previously produced commodities and services using existing productive resources. New technology has been an important source of *rising living standards* throughout U.S. economic history. A vast number of products or production processes now in general use were unavailable a century ago. Television, antibiotics, jet engines, atomic energy, computers, and many types of farm machinery are just some of many products now in use that were unavailable on a commercial scale before World War II. Polio vaccine was unavailable twenty-five years ago; hand-held calculators were unavailable ten years ago.

Since 1900, technological change has been the primary source of the rising living standard in this country. Many studies[1] attribute between one-third and one-half of the growth of real per capita income to technological change. Only the improved quality of the U.S. labor force (a development that is itself partly the result of the learning of new technology in schools and colleges) is responsible for a larger part of economic growth. Capital accumulation usually appears as the third major contributing factor, although its influence is often associated with technological change and vice versa.

A second benefit of technological progress is an *improved quality of living*. Some people think that technology merely helps produce more material goods which pollute the environment. On the contrary, technology has been, and can be, employed to abate pollution, conserve resources, improve education, and bring artistic performances to large numbers of people.

A third benefit is *greater national security*. Since World War II, the United States has relied on its lead in military technology to maintain military superiority over potential adversaries who have the advantages of numbers and low military personnel costs. Equally significant, technology has become important in peace-keeping endeavors, for example, in the surveillance of arms limitation agreements.

A fourth benefit is a *reduced rate of inflation as a result of increased productivity*. In general, inflation occurs when demand for goods and services outstrips supply; technological change can reduce inflationary pressures by raising productivity and thereby increasing the output produced by given resources. In earlier decades, some labor leaders and economists voiced the fear that rapid technological change would lead to large-scale technological unemployment. But these dire predictions have not been borne out, and such fears have subsided. In fact, studies show that employment tends to grow more rapidly in more technologically progressive industries because innovative activity involves additional expenditure for capital and labor. In addition, commercialization of innovation eventually leads to decreased costs that permit increased production and therefore increased employment.[2]

A fifth benefit is the assistance technological change provides in obtaining a *favorable balance of trade*. For the same reason that technological

---

[1] A variety of scholarly studies, many of them inspired by Edward Denison's comprehensive 1962 study of U.S. economic growth, which was the first of its kind, have cast light on this question. See Edward F. Denison, *The Sources of Economic Growth in the United States and the Alternatives Before Us* (New York: Committee for Economic Development, 1962).

---

[2] Roger Brinner and Mirian Alexander, *The Role of High-Technology Industries in Economic Growth* (Cambridge, Mass.: Data Resources, March 1977).

progress permits expanded domestic consumption, it also permits expanded foreign consumption and therefore expanded exports. Technological change encourages new or improved products and keeps prices down, thus making U.S. goods more competitive with those of foreign producers. Government data indicate that since the 1960s, the U.S. balance of trade has been better in technology-intensive products than it has been in other products.[3]

## SOURCES OF TECHNOLOGICAL PROGRESS

While independent inventors have traditionally been associated with developing new technologies, technological progress also emerges as people work with and improve the best existing technology. An engineer examining blueprints may perceive a more efficient way to link two processes. A worker who operates a complex industrial machine may make suggestions to improve the next generation of such machines. The history of shipbuilding offers an excellent example of such learning by doing. For hundreds of years, ships improved steadily, although no one undertook organized research on their design or construction.

In addition, during the last half century, organized research and development has grown and become an important contributor to technological change. This activity is carried on by a diverse set of institutions, including governments and universities, as well as businesses. Furthermore, research and development is itself a wide ranging set of activities lodged within an even much broader set of processes that bring new ideas from conception to the marketplace.

## PHASES OF TECHNOLOGICAL PROGRESS

The workings of technological progress can best be understood in terms of five related phases. Of course, all five phases are not necessarily present or distinguishable in every example of technological change. Similarly, these phases should not be regarded as constituting a sequence through which every technological change progresses.

The first and most elemental phase is *basic research*. It encompasses studies of the fundamental elements and processes of the universe. Typically, the motive of basic research is to produce knowledge for its own sake, without serious regard for the possibilities of useful application.

The second phase is *applied research*, in which the research and engi-

---

[3] National Science Board, *Science Indicators, 1976* (Washington, D.C.: National Science Foundation, 1977), Table 1–23. Technological intensity is measured by the relative amount of research and development performed by the pertinent businesses.

neering strives to apply basic knowledge to the solution of some particular problem or need. For example, applied research in atomic energy is built on the results of basic research in physics, and applied research in chemical engineering is built on basic research in chemistry. Often, the dividing line between these two phases is more an intellectual exercise than a practical division; in reality, laboratory work flows from one successful (or unsuccessful) experiment to another.

Once an applicable idea is proven in a laboratory setting, it still must go through testing and refinement in the third stage, *development*, to determine its commercial practicality. This phase includes the construction of pilot models and demonstration plants, as well as any related feasibility studies management may call for.

The combination of these first three phases is popularly labeled *R&D* (research and development). The general label *invention* is also applied to these activities. They have drawn a great deal of attention for scientific reasons, and many efforts have been made to measure their costs and benefits. But if technological progress stopped with these functions, society would gain comparatively little from it.

Realizing the fruits of invention requires a fourth phase in which it is incorporated into a full-scale producing plant. Moreover, this first-of-a-kind plant (called a *pioneer plant*) must be supported by capital investment, access to raw materials, labor, power, marketing facilities, and of course, consumer demand for the output. The sum total of all these actions is termed *innovation*.

The fifth phase of technological progress is the *diffusion* of the innovation throughout the economy. This final stage consists of replicating in a succession of other plants and firms the products and processes that have proved successful in a pioneer plant. How fast such diffusion occurs will depend on such factors as market receptivity, competitive conditions, the age of existing capital stock, and the overall pace of economic activity.

## RISKS AND REWARDS

Inventive activity is economically very risky both to the researcher and to the sponsoring organization. Research may or may not produce a useful idea, lead to new technology that can be patented, or lead to new technology that can be exploited by the institution that undertook the research. Successful research may lead to a minor improvement, to an unimportant product, or to an entirely new industry. New technology may or may not lead to a new product that can be produced, sold, and used in an economic, safe, and reliable fashion. Finally, the new product may or may not be acceptable to customers, and it may be displaced by yet another new product. Business experience and economic studies indicate that there is more uncertainty in the R & D phases than in fixed capital investment in plants that use the new technology. Normally, technological and feasibility risks decline as activity proceeds toward implementation through commercialization, although they are still considerable in the building of a first-of-a-kind commercial-size plant. They decline in the diffusion phase as duplicate plants are built.

On the other hand, the total dollar capital investment needed to bring new technology to the marketplace normally entails a much larger commitment of resources than does the research and development that generates the new technology. Thus, the combination of still considerable risks and a relatively large dollar requirement may be enough to deter the building of a pioneer plant, even after a technically successful R & D effort. Furthermore, duplicate plants using such new technology may not be constructed because market conditions cannot provide a competitive return on the needed capital investment.

To counter all these risks, society relies primarily on the potential of corresponding economic rewards. The market system rewards technological activities (and all others) by making payments to the holders of the ownership rights.[4]

Typically, however, ownership rights to the new knowledge that results from research and development are incomplete. The results of most basic research cannot be patented; applied research and development can result in ownership rights *if* the resulting new knowledge can be embodied in a patentable product or production process. When ownership rights to new technology are incomplete, there is little incentive for the private sector to undertake costly research and development.[5]

---

[4] In a competitive market economy, the system of private ownership or property rights (i.e., the rights to buy and to sell and the right to use property in the interests of the owner) is crucial to market performance. Workers have ownership rights to the labor they sell, producers have ownership rights to the inputs they buy and to the products they produce, and consumers have property rights to the products and services they pruchase. Such rights are essential in motivating the economic use of resources for high-priority purposes.

---

[5] Economists have tested this conclusion by attempting to measure social returns on research and development. Such attempts are fraught with difficulties especially in distinguishing between returns on research and development and returns on capital investment in which the resulting technology is incorporated. Nevertheless, most studies conclude that the average social return on research and development is in the range of 35 to 50 percent a year, which is much higher than the rate of return on fixed capital investment. The implication is that society would benefit from increased research and development. See National Science Foundation, Office of Economic and Manpower Studies, *A Review of the Relationship Between Research & Development and Economic Growth/Productivity*, November 1971, for surveys.

Any subsequent building of a pioneer plant is similarly inhibited if ownership rights to the technology are incomplete. But in the case of a product that is responsive to strong market demands, clear (even temporary) rights to exclusive production can yield a return that more than offsets the considerable risks facing the first builder-operator.

For subsequent plants that duplicate the new product or process, the rewards, like the risks, are more routine. In this final phase of technological progress, rewards are likely to depend primarily on the participants' operating efficiency, the strength of their particular markets, and such general factors as the state of the economy and the burden of taxation.

## FINANCING TECHNOLOGICAL PROGRESS

In most countries, basic research is undertaken largely outside the profit-making sector, primarily in universities and government laboratories; whereas applied research and development is undertaken on substantial scales by profit-making firms, by the private nonprofit sector, and by governments.

In the United States, the federal government finances just over half of all R & D activity but actually performs only 15 percent (see Figure 4). Industry and universities perform large amounts of federally funded research and development. Industry also finances substantial amounts of the research

**FIGURE 4**

**Performance and Financing of Research and Development, 1976**

(percent)

| Sector | Financed by | Performed by |
|---|---|---|
| Federal government | 53 | 15 |
| Industry | 43 | 70 |
| Universities and colleges | 2 | 10 |
| Federal R & D centers administered by universities | — | 3 |
| Other nonprofit institutions | 2 | 2 |
| Total | 100 | 100 |

SOURCE: National Science Board, *Science Indicators, 1976* (Washington, D.C.: National Science Foundation, 1977), Tables 2-4 and 2-5.

and development performed by universities. In addition, some 3 percent of research and development is performed by about twenty R & D centers specializing in national security and energy work funded by the federal government and administered by universities.

In 1976, U.S. expenditures for applied research were nearly twice those for basic research, and expenditures for development were nearly three times those for applied research. About two-thirds of basic research is financed by the federal government, and most of the remainder is divided about evenly between industry and universities. Just over half of applied research is financed by the federal government, largely in defense, space, health, and energy; most of the remainder is financed by industry. Financing of development activities is divided about evenly between the federal government and industry, with other sectors financing very little.

Most federal R & D support can be divided into three categories: activities in which the federal government is the principal user of the resulting new technology, activities in which the federal government has been given a broad mandate to promote science and technology, and broad support of basic research. Most funds in the first category go to defense and space research and development, which together use more than 60 percent of federal R & D funds. Health and environmental research and development are the largest uses in the second category; together, they account for about 20 percent of federal R & D funds. Broad-based support of science takes about 4 percent of federal R & D expenditures. Much of the remainder goes to energy research and development.

In the innovation and diffusion phases, the role of the federal government diminishes sharply. The government finances a relatively small number of production facilities; it actively operates even fewer, and those facilities are chiefly for improving defense.

Private industry dominates the production phase. It finances most of the new-technology plants and operates virtually all of them. Corporate expenditures for new-technology plant and equipment typically far exceed expenditures for research and development. The major cost of achieving technological progress therefore accrues after the R & D phase.

## TECHNOLOGICAL PROGRESS: A COORDINATED APPROACH

Although specific phases of technological progress can be identified, it is important not to isolate one phase from another. Decision makers at each point look both forward and backward, conditioning their own actions by their assessment of future markets and by the potential for economic rewards to be gained from new knowledge and improved techniques. The pull of market demand in the later phases of technological progress is therefore extremely important in stimulating investment in innovations and research and development. In a market economy, opportunities for sales to be gained by a new product (or threats of sales to be lost without one) are strong motivating forces that affect the entire process of technological advance.

Policy makers should realize that the close interdependence of the phases in the process argues against choosing one or a few points for *exclusive* policy attention. In some circumstances, lack of technological progress may be due to policies that produce disincentives or that fail to provide the necessary incentives at any one of the phases of the process. In such cases, policy reforms should give special attention to that particular problem area.*

---

*See memorandum by HENRY B. SCHACHT, page 72.

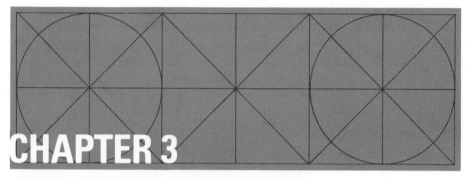

# CHAPTER 3
# THE ECONOMY AND FUTURE TECHNOLOGICAL PROGRESS

In the last decade, the performance of the U.S. economy has badly deteriorated. Although the causes and consequences of this deterioration, of course, extend beyond fixed capital investment, research and development, and innovation, the recent economic climate has been closely related to the issues of technological change and innovation.

## ECONOMIC CLIMATE AND DISINCENTIVES TO INNOVATION

Inflation increased from a 2.1 percent annual average rate in the 1958–1968 period to a 6.5 percent average in the most recent decade, climbing into the double-digit range in 1974 and again in 1979. Inevitably, sluggishness in the economy has caused the unemployment rate to rise gradually from 3.5 percent in 1968 to 6.0 percent in 1978. The annual growth rate of labor productivity in the private sector, which averaged about 3 percent from 1948 to 1965, fell to about 2 percent in the period 1966 to 1973 and on average has remained below 1 percent since 1973. Largely as a consequence, real income has grown slowly during the last decade, and real GNP has grown only about 2.8 percent a year, compared with 4.5 percent a year in the previous decade.

The economy has been not only sluggish but also erratic during the 1970s. We have had two recessions, and that of 1974–1975 was the most severe since the 1930s. Inflation has been ever present but variable, the price index having risen as little as 3 percent and as much as 11 percent from one year to the next. In addition, government policies have been unpredictable, running the gamut of price freezes, controls, decontrols, and now voluntary guidelines.

Business investment, which is necessary to ensure continued technological progress, rose only 2.6 percent a year in real terms from 1968 to 1978, compared with 6.2 percent a year during the preceding decade. Figure 3 (page 6) shows that the real rate of return on investment has declined dramatically since the mid-1960s. The return on total capital investment, which includes interest payments to corporate debt holders, peaked at almost 10 percent in 1965 and is currently barely over 5 percent. If only the return on equity (the return on the portion of capital stock financed by equity through new issues and retained earnings) is considered, the decline has been even more dramatic, with a current return of less than 5 percent.

Lack of a competitive rate of return on industrial investment has created a disincentive to the renewal and improvement of the nation's industrial plant and equipment, compared with the return on other investments. *Unless this disincentive is removed, it will be difficult to stimulate technological progress and overcome the long-term problems of inflation and lack of productivity.*

## POLICIES TO REVITALIZE TECHNOLOGICAL PROGRESS

Developing a broad range of policies that affect all phases of the technological process is one approach to stimulating technological change. Alternatively, it may be more appropriate to be selective and place high priority on reforming key policies. The choice of which policy changes should be emphasized obviously depends on which phases of the process of technological change are performing least effectively.

### IMPROVING THE CLIMATE FOR INVESTMENT AND INNOVATION

The ability of the U.S. economy to increase productivity has been less than that of America's major industrial competitors. The data in Figure 5 show that over the past decade, the increase in GDP per employed person has been much more rapid in all the major competing countries. And although rapid increases in the labor force participation rates of demographic groups with little work experience may have contributed slightly to the poor relative rate of growth within the United States, there must be more significant reasons why the trend in GDP per employed person has been seven times more rapid in Japan and about four times greater in Germany and France.

The level of GDP per employee is still higher in the United States than it is in the major industrial nations. Nevertheless, over the past decade, all countries except the United Kingdom have moved much closer to the U.S. level (see Figure 6). In Japan, the level of GDP per employee was about one-third of the U.S. level in 1967 and almost two-thirds of the U.S. level in 1978. France and Germany have now narrowed the gap between themselves and the United States to about 15 percent less than the U.S. level of GDP per employed person.

It is clear that America's major industrial competitors have improved their output more rapidly by making greater strides in productivity improvement, especially in the manufacturing sector. In fact, there is some evidence to suggest that several major industrial countries, particularly Japan and

**FIGURE 5**

**Change in GDP per Employed Person, Selected Countries, 1967 to 1978**

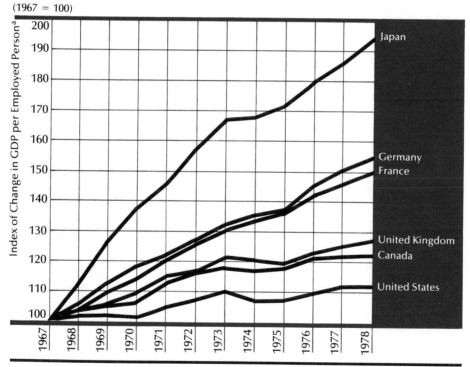

[a] Index is based on each country's own prices corrected for inflation. Consequently, the data show the trend in GDP per employed person within each country. They do *not* represent a comparison of the relative level of GDP per employed person among countries.

SOURCE: U.S. Department of Labor, Bureau of Labor Statistics.

Germany, now have equal or higher *levels* of productivity in some manufacturing industries than the United States.[1]

For all industrial countries, improvement in manufacturing productivity fell in the 1970s in comparison with the 1960s (see Figure 7). However, the U.S. rate of improvement has been consistently lower than that of all other major industrial countries except the United Kingdom. *Over the past two decades, U.S. manufacturing productivity growth has ranged between one-third and one-half of the rate for Japan, Germany, and France.*

Productivity growth obviously varies among industries and sectors of the economy. In the United States, there has been a significant decline in productivity growth rates in most major sectors of the economy for the pe-

---

[1] Levels of productivity obviously vary among industries within manufacturing. A 1972 study by Yukizawa found that Japanese productivity exceeded U.S. productivity per employee for sixteen manufacturing products out of sixty studied. See Angus Maddison, "The Long-run Dynamics of Productivity Growth," *Banca Nazionale del Lavoro Quarterly Review*, Rome, March 1979, p. 11.

---

**FIGURE 6**

**Comparison of Real GDP per Employed Person in United States and its Major Industrial Competitors[a]**

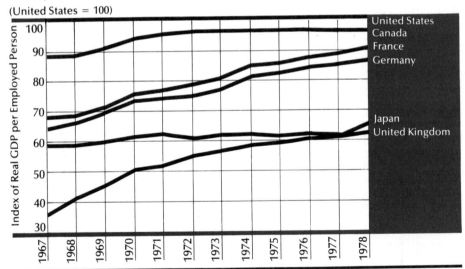

[a] Output in each country has been converted to U.S. prices. The chart therefore shows a comparison of the relative level of GDP per employed person among countries. If international price weights are used, the level of productivity for countries other than the United States would be several percentage points lower than shown.

SOURCE: U.S. Department of Labor, Bureau of Labor Statistics.

## FIGURE 7

**Capital Investment and Output per Hour, Selected Countries***

| | Capital Investment as Percentage of Output** | | | | | Percentage Annual Growth in Output per Hour*** | |
|---|---|---|---|---|---|---|---|
| | Total Economy | | Manufacturing | | | Manufacturing | |
| | 1960–1969 | 1970–1977 | 1960–1969 | 1970–1977 | | 1960–1970 | 1970–1978 |
| United States | 14.9 | 14.5 | 8.8 | 9.6 | | 2.9 | 2.7 |
| Canada | 20.0 | 19.3 | 14.4 | 15.1 | | 4.3 | 3.6 |
| Japan | 28.8 | 26.7 | 29.9 | 26.5[a] | | 10.8 | 5.0 |
| France | 19.5 | 18.8 | NA | NA | | 5.8 | 5.1 |
| Germany | 20.1 | 18.7 | 16.3 | 15.2[b] | | 5.5 | 5.3 |
| United Kingdom | 16.5 | 17.6 | 13.4 | 13.6 | | 3.5 | 2.1 |

*Capital investment excludes residential construction.
**For comparative purposes output is measured at current factor cost.
***All employed persons for U.S. and Canada; all employees for other countries.

[a] 1970–1974.
[b] 1970–1976.

NA = Not available

SOURCE: U.S. Department of Labor, Bureau of Labor Statistics.

riod 1973–1978 compared to growth rates in 1965–1973. The communications industry was an exception to this downward trend with its productivity growth rising rapidly to an average of 7.1 percent annual growth for 1973–1978. In mining a negative productivity growth rate for 1973–1978 actually lowered the level of productivity in the industry. Manufacturing productivity which was improving at a rate of 2.4 percent for 1965–1973 declined to a rate of 1.7 percent for 1973–1978. In finance, insurance, and real estate, productivity rose slightly to an annual average of 1.4 percent change and in services productivity growth declined to 0.5 percent in 1973–1978. However, in both these sectors quality changes in output are probably not adequately reflected in output measures and if there were more rapid quality improvements in these sectors in recent years, compared to quality improvements in the late sixties and early seventies, productivity growth in these industries would currently be understated.[2]

Many factors influence productivity, but unless the efficiency of plant and equipment is improved it is impossible to sustain a high rate of productivity growth. The United States, which was once in the forefront of using technological innovation to raise the productivity of its economy, no longer occupies that position in the industrial world. A major part of the explanation for this change is that the other major industrial countries have invested proportionately more of their resources in plant and equipment that embody technological innovations.

Figure 7 shows that for the total economy, capital investment as a percentage of output was substantially lower in the United States than it was in almost all other industrial countries in both the sixties and the seventies. Proportionally, U.S. investment was about two-thirds of French and German investment and about one-half of Japanese investment. While the U.S. rate of capital investment in the total economy was lower than the rate for other countries *the most significant difference between the United States and other countries has been our much lower rate of investment in the manufacturing sector.*[3]

---

[2] For a discussion of the decline in productivity in the United States, see J.R. Norsworthy, Michael J. Harper, and Kent Kunze, "The Slowdown in Productivity Growth: Analysis of Some Contributing Factors," *Brookings Papers on Economic Activity*, (Washington, D.C.: The Brookings Institution, forthcoming 1980).

[3] This conclusion is supported by data on real nonresidential capital investment during the 1970–1976 period. In each of these years, the gross addition to capital stock per capita, corrected for price changes and based on international price weights, was higher in France, Germany, and Japan. Since 1971, their gross additions each year have been between 15 to 30 percent higher than those of the United States. Data supplied by U.S. Department of Labor, Bureau of Labor Statistics. Please note information will be included which describes U.S. capital investment from 1976. These data will show that since 1976 the capital/labor ratio has been falling.

Lack of capital investment in the United States has resulted in a decline in the rate of change in the capital/labor ratio in the economy. For the private business sector the average annual rate of change in this ratio, which reflects the amount of plant and equipment labor is given to work with, was 2.21 percent for the period 1965–1973 but declined to 0.67 percent in 1973–1978. For manufacturing, the decline was from 2.49 percent in 1965–1973 to 1.89 percent in 1973–1978.[4] The increasing importance of non-manufacturing in the private business sector may account for some of the decline in the rate of change in the overall capital/labor ratio for the private business sector. However, *the decline in the rate of change for manufacturing is especially significant since in the past this sector has increasingly used capital intensive methods of production, frequently embodying new innovations, to raise productivity.* Available evidence suggests that inadequate capital investment especially in manufacturing has played an important role in the reduction of the overall U.S. rate of productivity improvement and that other countries are now challenging the preeminent position America once held in innovation in many industries.

This relative weakening of the competitive position of U.S. manufacturing is reflected in a declining share of total U.S. manufactured goods in world trade. For example, in 1971, the United States accounted for about 21 percent of the total manufactured goods exported by the world's fourteen major industrial countries. That share declined to a little over 20 percent in 1975 and dropped dramatically to about 16 percent at the beginning of 1978.[5]

Clearly, lack of capital investment has contributed substantially to poor U.S. productivity performance. Consequently, any attempt by the government to revitalize technological progress must include appropriate policies that will provide greater incentives for capital investment.

## IMPROVING THE CLIMATE FOR RESEARCH AND DEVELOPMENT

Over the past twelve years, U.S. expenditures for research and development have declined as a share of GNP. The data in Figure 8 show the extent of the decline in the proportion of R&D expenditures in the United States and the trends in several other industrial countries. As a proportion of GNP, nondefense R&D expenditures in the United States rose in the first half of the 1960s but then gradually declined. There has been an increase in the proportion of resources devoted to research and development in Ja-

---

[4] J.R. Norsworthy, Michael J. Harper, and Kent Kunze, "The Slowdown in Productivity Growth: Analysis of Some Contributing Factors," op. cit., Appendix Table 1.

[5] U.S. Department of Commerce, 1979, Bureau of International Economic Policy and Research.

## FIGURE 8

**R & D Expenditures as a Percent of GNP, Selected Countries, 1961 to 1977**

| YEAR | United States | | France | | Germany | | Japan | | United Kingdom | |
|---|---|---|---|---|---|---|---|---|---|---|
| | (1) | (2) | (1) | (2) | (1) | (2) | (1) | (2) | (1) | (2) |
| 1961 | 2.74 | 1.34 | 1.38 | 0.98 | NA | NA | 1.39 | 1.37 | 2.39 | 1.48 |
| 1967 | 2.91 | 1.87 | 2.13 | 1.59 | 1.97 | 1.81 | 1.53 | 1.52 | 2.33 | 1.68 |
| 1972 | 2.43 | 1.66 | 1.81 | 1.55 | 2.33 | 2.18 | 1.85 | 1.84 | 2.06 | 1.53 |
| 1975 | 2.30 | 1.62 | 1.82 | 1.39 | 2.39 | 2.23 | 1.94 | 1.93 | 2.05 | 1.52 |
| 1976 | 2.27 | 1.57 | 1.74 | 1.47 | 2.28 | 2.15 | NA | NA | NA | NA |
| 1977 | 2.25 | 1.59 | NA | NA | NA | NA | NA | NA | NA | NA |

NA: Not available.
(1) Total R & D expenditures.
(2) Total R & D expenditures (excluding national defense)

SOURCE: U.S. Department of Labor, Bureau of Labor Statistics, based on data from the National Science Foundation, April 1979.

pan and Germany, this country's major industrial competitors. Moreover, the rapid growth of GNP in those two countries increases the importance of their greater emphasis on research and development.

There is some concern that there may have been a weakening in the U.S. government's commitment to direct expenditures on research and development. This concern is based on the fact that the decline in R & D expenditure as a proportion of GNP has been entirely in government-financed projects. In fact, the industry and university proportions of the total R & D share of GNP have actually increased slightly since 1967.

The principles by which government R&D spending ought to be determined will be discussed in Chapter 7. Here, it only needs to be noted that most government R&D spending is for specific government needs. The largest share of R&D expenditure was devoted to defense and aerospace, both of which were sharply curtailed in the late 1960s and early 1970s. In recent years, however, defense research and development has been increasing rapidly.

It is certain that the market economy benefits either directly or indirectly from most kinds of federal R&D spending, but the extent of those benefits is a matter of judgment and controversy. Certainly, aerospace research and development had indirect beneficial effects on the market economy, and parts of the space program have yielded important direct market benefits in the fields of health, meteorology, and resource exploration. It is appropriate, of course, that mission-oriented federal R & D spending should be decreased to match any reduced needs of government agencies. However, if there is a reduction in mission-oriented government R & D and procurement which has had an important indirect effect on innovation of commercial products and services, there will be some slowdown in the rate of technological innovation unless other equally effective government or private efforts are initiated to offset such reductions. From the point of view of the market economy, the greatest benefits will be obtained if government spending for applied research and development is replaced by business spending, with government taking the lead in support of basic research.

This Committee believes that the available evidence suggests that increased expenditure on research and development would benefit technology progress and that there is need for special measures to stimulate such expenditures. *Our analysis of the current economic climate and its effect on technological change shows that a more favorable climate for business investment is the most vital ingredient needed to stimulate innovation and economic growth in the United States.* Greater expenditure on plant and equipment will result in innovations based on the existing output of past research and development and will also stimulate the diffusion of innovations throughout industry. This is essential for reversing the current unfavorable trend in productivity.

In addition, capital investment will encourage even greater business expenditure on research and development. It is impossible to disentangle the impact of research and development and fixed capital investment spending on economic growth in a precise and detailed way. In many cases, they are separate stages of a single long-term business investment program.

The decision to invest in new technology and in new capital is made by a company on the basis of a comparison of potential projects. Management makes an assessment of the risks involved, the profits to be gained, and the cost of carrying each project forward. Anticipated profits are, in turn, influenced by investor confidence in the future growth and stability of the economy. The more concerned investors are about stagnation and downturns, the less likely they will be to commit funds to long-term investment in research and development and to fixed capital investment. Similarly, the more uncertain the economic prospects are, the more probable it is that potentially important commercial innovations involving a large element of risk will be postponed or rejected. It follows that the declines in the level, rate, and quality of industrial innovation over the past decade can be attributed in part to the destabilizing effects of inflation, unpredictable and unproductive government regulation, the energy problem, and an uncertain energy policy response.

The government must initiate policies that will create incentives for business to invest in new plant and equipment. Such actions will demonstrate government's commitment to a high and rising level of investment. An effective government program to stimulate innovation and productivity growth must recognize that if firms are to undertake increased investment in innovation, there must be improved assurance of potential economic reward, much more stable prices, and hence, more cause for investors to have confidence that tomorrow's economy will reward today's efforts.

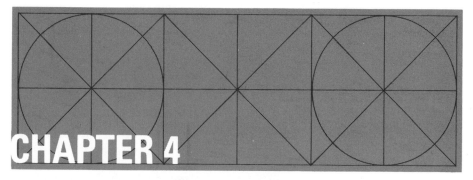

# CHAPTER 4
# TAX POLICIES FOR INVESTMENT AND INNOVATION

Government economic and fiscal policies influence the pace of private research and development and innovation in three distinguishable ways: through their influence on the broad economic environment in which the private sector is operating, through the impact of general tax measures on the profitability and cash flow of individual firms, and through the impact of selective tax incentives and disincentives aimed specifically at the research, development, and innovation activities of a firm.

This chapter reviews the effect of general and selective tax measures on innovation and research and development and concludes with a series of tax proposals that should be given individual policy consideration. On the basis of this review this Committee believes that the most effective way for policy makers to stimulate technological progress is to give top priority consideration to the following two tax changes:

- **The gradual introduction of a more rapid capital recovery allowance to stimulate investment in new plant and equipment and encourage research and development**
- **The introduction of flexible depreciation for R&D structures and equipment to assure up-to-date research facilities in industry**

## IMPACT OF GENERAL TAX MEASURES ON INVESTMENT AND INNOVATION

There is considerable evidence showing that general tax incentives stimulate increased business investment and thereby stimulate a faster pace of industrial invention and innovation.[1] Moreover, numerous studies of innovation have found that business liquidity and profitability, both of which are affected by tax policy, have a significant impact on innovative effort.[2] Because well over half of the total investment required for a successful innovation usually comes after the completion of the various R & D phases,[3] the case for concentration on the overall investment climate is particularly strong. *Consequently, future technological progress will depend heavily on general investment incentives for the overall economy.*

The government has powerful policy instruments at its disposal to affect corporate profitability and cash flow and thereby to affect the climate for investment and innovation. These include the federal corporate income and capital gains tax provisions and the regular depreciation allowances associated with those provisions.* When undertaken within the constraints of an appropriate overall fiscal policy, liberalization of such tax measures can increase business investment and innovation in three ways:

- by increasing rates of return on new projects (either new facilities or improvements in existing facilities) and thereby making more projects economically attractive

- by increasing the flow of funds from existing projects and thus enabling the firm to invest in more new projects

- by reducing the tax burden on the private sector

Moreover, increased business investment and innovation can, in turn, produce other positive effects: In an underemployed economy, more peo-

---

[1] See Jorgenson and Hall, Bischoff, and Coen in G. Fromm, editor, *Tax Incentives and Capital Spending* (Washington, D.C.: The Brookings Institution, 1971).

[2] D. C. Mueller, "The Firm Decision Process: An Econometric Investigation," *Quarterly Journal of Economics*, Vol. LXXXI, No. 1 (February 1967) p. 58; and H. G. Grabowski, "The Determinants of Industrial Research and Development: A Study of the Chemical, Drug, and Petroleum Industries," *Journal of Political Economy*, Vol. 76, No. 1 (March-April 1968) p. 292.

[3] Estimates of the cost ratio of R & D expenditures to innovation expenditures (commercialization) range from 1: 1.5 to as low as 1: 10. See Robert Gilpin, "Technology, Economic Growth and International Competitiveness," Joint Economic Committee, Congress of the United States (Washington, D.C.: U.S. Government Printing Office), July 9, 1975, p.6; and Edwin Mansfield, *Research and Innovation in the Modern Corporation* (New York, N.Y.: W.W. Norton, 1971), p. 118.

---

*See memorandum by RODERICK M. HILLS, page 69.

ple will be immediately put to work implementing the new investments; in a fully employed economy, output per person will be increased through productivity-enhancing investments in better machines, better processes, and better training in how to use them. In either case, the resulting higher real incomes lead to demand for more goods and services, which, in turn, requires further new product and productivity-enhancing investment.

Federal fiscal policy plays an important role in determining the overall health of the national economy. In view of the current serious inflation, restraint on the federal budget must be given high priority. But an appropriate fiscal policy also means holding down the overall government tax and expenditure burden on the private sector in order to allow the incentives and motivations of the market system to work effectively.[4] Today, in our judgment, that overall burden is higher than it ought to be.

A congressional report shows that despite recent federal tax cuts, tax revenues have risen steadily as a share of GNP.[5] Federal taxes were 18.5 percent of GNP in 1976, 19.7 percent in 1978, and are forecast to reach 20.3 percent by 1980 if the current tax law remains unchanged. Similarly, as the data in Figure 2 (page 5) show, when profits are adjusted for inflation, it is evident that the effective tax rate on corporate profits has risen steadily from 1977 to 1979.

Government officials face the task of achieving a judicious trimming of both taxes and expenditures and to make sure that the timing and amounts of such changes are consistent with a responsible overall budget. Therefore, we urge the government to give strong weight to the powerful effects of certain general tax measures on stimulating technological progress. Among the general tax changes that should be considered are a more rapid capital recovery allowance, regular rate reductions in income taxes, and a reduction in the capital gains tax.

**This Committee concludes that a more rapid capital recovery allowance is the first-priority action among *all* the alternative tax measures.** We have concluded that the high and uncertain rate of inflation is a critical barrier to investment in the longer-term ventures that are characteristic of technological innovation. We believe a long-run supply-oriented attack on inflation should be made by stimulating capital investment and productivity growth. More capital investment would bring about a higher rate of diffusion of advanced technology and would encourage more investment in

---

[4] For a fuller discussion of these concerns, see *Redefining Government's Role in the Market System*.

[5] *Toward a Balanced Budget*, Committee on the Budget, U.S. House of Representatives, April 13, 1979.

technology development. Faster capital recovery rates for new plant and equipment would be a more effective stimulus to technological innovation and a more efficient approach for this purpose than other potential tax changes.

## A MORE RAPID CAPITAL RECOVERY ALLOWANCE

If business were permitted to depreciate capital plant and equipment over fewer years than the current tax law stipulates, the existing capital recovery allowance would be improved. Increased recovery allowances immediately raise cash flow and effectively reduce the cost of investment.

Figure 9 shows the relationship between cash flow and expenditures on plant and equipment and on research and development for U.S. nonfinancial corporations. Although there are variations, the trend line shows that plant and equipment expenditure is generally associated with cash flow. There are many possible explanations for the variation that seems to have developed in the mid-1970s. For example, the relative attractive-

**FIGURE 9**

**Relationship between Expenditures on Research and Development and on Plant and Equipment and Cash Flow, U.S. Nonfinancial Corporations, 1960 to 1978**

(1960 = 100)

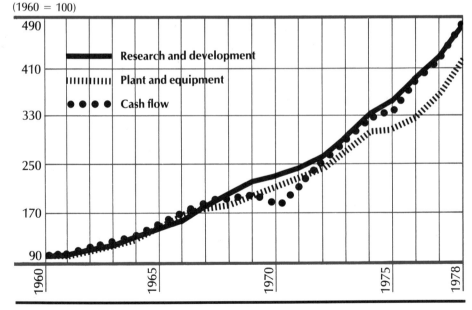

SOURCE: U.S. Department of Commerce, Bureau of Economic Analysis.

ness of investment in nonfinancial corporations (and plant and equipment) may have declined compared to other investment opportunities. Rising energy prices may have increased the need for working capital and reduced the cash flow available for investment in plant and equipment. Finally, in an inflationary period, the increasing discrepancy between allowed capital recovery rates and replacement costs has a detrimental effect on capital investment. The present tax system discourages investment in new equipment and machinery because allowable depreciation rates are based on original costs, even though replacement costs are much higher as a result of inflation during the life of the asset.

Last year, the Commerce Department estimated that the difference in pretax profits as a result of basing depreciation allowances on historical rather than replacement costs was $18 billion. At a 48 percent marginal tax rate, corporations paid almost $9 billion in direct taxes on these illusory profits; and a recent National Bureau of Economic Research study suggests that when indirect effects are taken into consideration, the total added tax was several times greater.

One way to eliminate this "inflation tax" on capital would be to index depreciation allowances for rising replacement costs resulting from inflation. A less administratively burdensome method would be a more rapid tax write-off of depreciable investments. Such a change would accelerate cash flow and would reduce investor vulnerability to technological obsolescence and to inflation by allowing recovery of investment dollars of more equal purchasing value than is now possible. Such a change *defers*, rather than eliminates, investor tax liability, an important consideration to keep in mind in debates over alternative tax changes.

**This Committee therefore concludes that the benefits of phased introduction of accelerated depreciation are such that some form of more rapid capital recovery is urgently required.** With a more rapid recovery allowance for new plant and equipment, there is greater incentive for business to invest in new plant and equipment in order to gain the benefit of the incentive and at the same time provide an incentive toward longer-term investments. This type of tax policy change would immediately stimulate the introduction of innovations embodied in new plant and equipment and **this Committee places highest priority on this recommendation.***

A change in the capital recovery allowance will result in an initial revenue loss for the U.S. Treasury. However, we urge that as policy makers consider changing the tax code, strong emphasis be placed on the relatively powerful productivity- and income-creating potential of the faster capital recovery allowance that we have recommended. Those effects are difficult

---

*See memorandum by ROBERT R. NATHAN, page 72.

to estimate, and they are not immediate; nevertheless, they are substantial and will offset the initial revenue loss.[6]

Other types of tax changes that could be considered for stimulating technological progress when the government's fiscal priorities permit are discussed on pages 40 and 41.

## IMPACT OF SPECIAL TAX MEASURES ON RESEARCH, DEVELOPMENT, AND INNOVATION

There is a strong correlation between the growth of industry sales and investment and industry-financed R & D expenditures.[7] Healthy rates of business expansion and growth in market demand encourage more industrial research and development.

The broad tax reductions we have proposed would help remove the disincentives to research, development, and innovation resulting from high rates of taxation of property income and high rates of inflation. Although R & D investment would be enhanced by a more rapid capital recovery allowance, deficiencies in R & D expenditures would remain because of imperfect ownership rights to new technology.[8] Therefore, there is a need for

---

[6] One indication of potential impact was recently demonstrated in an econometric simulation by Data Resources with respect to the dynamic feedback effects of accelerated depreciation. They estimated that a reduction in depreciation lives to five years for equipment and ten years for nonresidential structures, using sum-of-the-years-digits depreciation, would by the third year raise business capital investment by $30 billion, GNP by $54 billion, productivity growth by 0.4 percentage points a year, employment by 1.5 million, and Treasury tax receipts by $8 billion. In addition, the unemployment rate would be 1.2 percentage points lower than 1981. Although different assumptions may lead to variations in results, these calculations point in the proper direction: to a sizable benefit over time from these changes. See Data Resources, *Tax Policy, Investment and Economic Growth*, prepared by Securities Industry Association based on Econometric Studies by Data Resources, March 1978. Different time periods for depreciation and the speed with which more rapid depreciation is phased in will, of course, lead to a different effect on GNP, unemployment, productivity growth, inflation, and Treasury receipts. For example, an unpublished Data Resources study (June 1979) provides estimates different from those quoted here and shows tax receipts actually declining.

[7] See discussion by Robert Gilpin, "Technology, Economic Growth and International Competitiveness." Joint Economic Committee, Congress of the United States, (Washington, D.C.: U.S. Government Printing Office), July 9, 1975, p. 35; research by K. Pavitt and W. Walker, "Government Policies Toward Industrial Innovation: The Final Report of a One-Year Study" (The Four Countries Project). Vol. A: Main Text and References. November 8, 1975, p. 5; and J. Schmookler, *Invention and Economic Growth* (Cambridge, Mass.: Harvard University Press, 1966).

[8] There is, however, little evidence that shows whether specific R & D cost-reducing tax incentives *alone* could induce more invention and innovation. This is partly because substantial R & D incentives have rarely been tried in the United States. Attempts to measure the effectiveness of foreign R & D tax incentives on inventive and innovative effort have been largely unable to distinguish tax effects from other potential influences. See OECD, *The Conditions for Success in Technological Innovation* (Paris, 1971); R. S. Kaplan, "Tax Policies of U.S. and Foreign Nations in Support of R & D and Innovation," Chapter 1: *Tax Policies for R & D and Technological Innovation* (Pittsburgh: Carnegie-Mellon University, 1976).

selective policies and programs to increase the profitability of R & D investment *relative* to other uses of productive resources. That relative increase can, in turn, generate technological impetus that will reinforce the pull of market demand for new and improved products by calling forth greater investment in innovation.

We believe that selective tax measures would be a good supplement to the general tax change we recommend. Such measures can be designed to involve small total tax revenue changes and therefore can be accommodated within a responsible overall fiscal policy. Little information exists on the effectiveness of selective tax incentives, compared with general tax changes, in inducing invention and innovation. But in the Committee's view, such selective incentives are most likely to be productive in improving the overall environment for investment when they are introduced along with a program of general tax adjustments (rather than separately).

In brief, the Committee feels that the combination of selective and general tax incentives can stimulate more private investment in invention and innovation by lowering its cost relative to that of other endeavors, by increasing the potential rewards, by decreasing the risk or cost of commercial failure, and by increasing the supply of investment funds.

Government could adopt a number of programs to try to increase the relative profitability of commercial research and development. **This Committee believes that, on balance, an additional selective tax incentive combined with general tax measures already recommended are the most desirable measures with which to encourage research and development.**

Unlike measures such as federal loan guarantees, government procurement of high-technology goods, and government equity financing of high-risk technological innovation, tax incentives do not require that government officials make difficult, subjective judgments concerning the relative merits of various innovations and technologies. Only the conviction of the individual inventor or innovator counts in the qualification for broad-based, nondiscriminatory tax incentives.

Another advantage of tax incentives is that they do not create artificial markets. Firms are still free to design, price, and sell in response to real demand, rather than in response to government-created demand. Because they provide one-time (rather than ongoing) financial support, tax incentives offer no assurances of later government relief if a technology fails or a new product does not win consumer acceptance. Tax incentives work with the market, not against it, in penalizing and eliminating inefficient procedures and managements and in encouraging and promoting efficient ones.

Compared with foreign tax policies, the U.S. tax system offers little encouragement for industrial research and development and only standard investment incentives for innovation. "Special U.S. Tax Provisions Affecting Invention and Innovation" (below) provides a brief summary of the U.S. situation. The more assiduous attention paid to this subject by America's major trade competitors is summarized in "Foreign versus U.S. Tax Treatment of Invention and Innovation" (page 38).

## TOWARD BETTER SELECTIVE TAX INCENTIVES

Why has the United States resisted such incentives? In part, it may be that American technological leadership seemed so strong and so secure that little attention was given to preserving its sources from erosion.

Resistance has also arisen because of belief that such incentives will aid the wrong activities, will not be cost-effective, or will benefit less deserving taxpayers. Most existing and proposed tax incentives do little to discriminate between firms that are highly successful in producing useful inventions

---

**SPECIAL U.S. TAX PROVISIONS AFFECTING INVENTION AND INNOVATION**

A few sections of the U.S. Internal Revenue Code single out the *invention* of new technical processes and know-how for special consideration. However, the code always treats *innovation*, or the commercialization of such invention in the marketplace, in the same way it treats other business endeavors.

IMMEDIATE EXPENSING OF RESEARCH AND DEVELOPMENT Section 174 of the U.S. Internal Revenue Code gives all business taxpayers the option of deducting research and experimental (development) expenditures in the year they are incurred, even though such expenditures may lead to the creation of an intangible asset (such as a patent or unpatented know-how) with a useful life in excess of one year. Like accelerated depreciation of tangible business investments, the instantaneous write-off of R & D expenditures aids business by deferring tax liability and increasing cash flow. The longer the life of the assets created, the higher the time cost of money; and the greater the proportion of R & D outlays to the total costs of invention and innovation, the more important this stimulus becomes.

Section 174 also provides encouragement for the taxpayer in the determination of R & D expenditures. Instead of the usual test that deductible expenditures be incurred "in carrying on" an existing trade or business, R & D outlays need only be incurred "in connection with" a current trade or business. Independent inventors and small businesses with a single or very few product lines are the principal beneficiaries of this

and innovations and firms that are unsuccessful. These incentives tend to reward inventive effort (R & D expenditures), rather than commercial success. To a large extent, this is unavoidable because of the difficulty of identifying what is an invention or innovation deserving of public support. What, for example, are suitable criteria for establishing whether enterprises are technologically progressive or involved in high-risk or long-term scientific projects? Moreover, how are the revenues and profits from desirable invention and innovation to be distinguished from overall profits?

Uncertainty is pervasive in industrial invention and innovation, and forecasts of the future success of individual R & D projects are notoriously unreliable. **Consequently, we believe it would be unwise to restrict tax benefits to specially defined recipients.** Restrictions are likely to cost more in lost opportunities (and potential tax abuse) than they initially seem to gain in conserved tax revenues. The firms with the most innovative successes are

---

more liberal interpretation of deductibility. Unlike large, diversified companies, whose R & D efforts are commonly pursued "in carrying on" existing commercial ventures, smaller firms often carry on very little commercial activity that is related to their technological research and experimentation. Sometimes, as in the case of small professional inventors, they carry on very little commercial activity at all other than their pursuit of salable inventions. Hence, they must rely on the broader "in connection with" test to achieve deductibility of expenses in the search for new products and inventions.

PATENTS Section 1235 of the Internal Revenue Code provides capital gains treatment for the sale of patent rights by the inventor or any other individuals who purchased the undivided rights to an invention before it was put to practical use. This constitutes a tax advantage because the patent in the hands of the inventor might otherwise be considered stock in trade (rather than a capital asset), with the result that income from its sale or exchange would be taxed at ordinary income rates. In contrast, the code does not specifically provide the benefits of capital gains treatment in the case of the sale of copyrights by an author.

But Section 1235 does not apply to corporations or employers who hold the rights to patents developed by their employees. Such employers must rely on general tax rules and case law to obtain capital gains treatment on the sale of patents. Among other things, this means that the sale of employer-owned patents will result in ordinary income tax treatment unless the employer has held the patent for more than one year in a manner that cannot be considered the holding of business inventory or stock in trade.

usually those that spend the most on research and development. Studies of industrial innovation have consistently revealed a high correlation between levels of R & D expenditures and results such as patents, productivity growth, and revenues generated by new products.[9] Therefore, encouraging expenditures appears to be the most practical method of encouraging the desired results.

## REDUCING THE COST OF INVENTION AND INNOVATION

The Committee believes that a number of selective tax reforms designed to reduce the cost of industrial invention and innovation could be considered by policy makers.

---

[9]/ See, for example, W. N. Leonard, "Research and Development in Industrial Growth," *Journal of Political Economy*, Vol. 79, No. 2 (March-April, 1971), p. 232; N. Terleckyj, *The Effects of R & D on Productivity Growth of Industries: An Exploratory Study* (Washington, D.C.: National Planning Association, 1974); and Z. Griliches, "Research Expenditure and Growth Accounting," in *Science and Technology in Economic Growth*, B. R. William, ed. (New York: Halsted Press, 1973).

---

### FOREIGN VERSUS U.S. TAX TREATMENT OF INVENTION AND INNOVATION

The U.S. tax code offers no special relief for innovators who incur the heavy costs of either the design, construction, and tooling of manufacturing facilities to produce the innovation for the first time on a commercial scale or the manufacturing and marketing start-up costs that assure the diffusion of the innovation into the economy. Moreover, the United States offers no special incentives, beyond those available for investment generally, for investment in depreciable structures and equipment used for research and experimental activities.

This is not the case in numerous foreign nations.[1] The governments of Japan, West Germany, and France, for example, have provided special accelerated depreciation allowances to approved investments in new technology (Japan) or to plant and equipment devoted to research and development (Germany) or used for scientific and technical research (France). In Japan, companies that form joint research associations in certain industries can immediately expense the cost of contributions for new machinery and equipment or a new facility. In Germany, a 7.5 percent tax-free cash subsidy is provided for investment in R & D facilities; and in France, special companies formed to perform research and development or apply innovative processes receive highly favored tax treatment on all kinds of activity.

---

[1]/ This survey of foreign tax provisions is largely based on R. S. Kaplan, "Tax Policies of U.S. and Foreign Nations in Support of R & D and Innovation," Chapter 1, *Tax Policies for R & D and Technological Innovation* (Pittsburgh: Carnegie-Mellon University, 1978), pp. 34–37.

In addition to the recent decline in overall R & D expenditures as a proportion of GNP, the rapidly rising cost of maintaining an efficient and progressive level of R & D effort in industry with up-to-date facilities and instrumentation is of major concern. The rapidly increasing level of sophistication in facilities required for the conduct of modern research and development means that such facilities become obsolete more rapidly than plant and equipment used directly in production. **Consequently, this Committee recommends that the most effective way of stimulating R & D investment is to change current tax policy for depreciating R & D structures and equipment.** This change would ease the so-called inflation tax on earnings from capital invested in research and development.

The tax code currently allows a deduction for depreciation allowances for buildings and other depreciable property, including equipment, used in R & D activities. The code should be amended to allow *flexible* depreciation for all such fixed assets. Under such a system, the taxpayer would have the

---

Like the government of the United States, the governments of Japan, West Germany, France, and almost all other major industrialized nations allow current deductibility of R & D expenditures, but Japan makes these expenditures even less costly by providing a 20 percent tax credit expense that represents an increase over the highest R & D expenses incurred by the company in any year after 1965. In Germany, an equally novel arrangement provides that, in some cases, an individual's supplementary income from scientific activities is taxed at half the normal rate. And in France, sales of patent rights, technical and manufacturing processes, and know-how are taxable at the sharply reduced 15 percent long-term capital gains rate.

It should be noted that in addition to these special provisions, many industrial nations give considerably more favorable treatment in general to capital gains and depreciation allowances than the United States does.

It is clear that foreign nations provide more generous tax incentives for invention and innovation than the United States does. Whether this leaves the United States at a serious competitive disadvantage in international markets is an issue too complex to be resolved here. However, it should be noted that a number of recent analyses have perceived slippage in the traditional worldwide leadership of the United States in the production of R & D–intensive goods.[2]

---

2/ See National Research Council, *Technology, Trade, and the U.S. Economy* (Washington, D.C.: National Academy of Sciences, 1978).

option of depreciating these assets fully in the first year of their life or adopting any other method desired while retaining the benefits of the allowable investment tax credit. This flexible depreciation would introduce incentives into the U.S. tax system similar to those that exist in a number of major foreign economies, including Canada and the United Kingdom.

## ADDITIONAL TAX CHANGES FOR POLICY CONSIDERATION

In addition to the two high-priority tax policy recommendations for improving capital recovery allowances and flexible depreciation for R & D equipment and buildings, we believe that policy makers should review a number of other potential improvements in tax policy that can play an important role in stimulating technological progress.

The Analysis supporting these recommendations is presented in Appendix A.

## GENERAL INCENTIVES

REGULAR RATE REDUCTIONS ON INCOME TAXES Gradual across-the-board rate reductions in federal taxation on all sources of individual and corporate income would increase both the capacity and the incentive for investment. Rate reductions would provide an important signal to innovators that government is serious about restraining the growth in federal expenditures, encouraging the private sector, and correcting the nation's productivity slowdown.

REDUCTION IN THE TAX ON CAPITAL GAINS Reducing the capital gains tax will encourage investors to accept risks in the hope of receiving greater rewards for innovative success. While this Committee endorses the 1978 federal tax reductions in long-term capital gain taxes, a further permanent cut would be particularly beneficial to small, technology-based firms that are heavily dependent on equity to finance innovative activities.[10] A constructive companion reform would be tax-free treatment of capital gains that are reinvested.

REDUCTION IN MAXIMUM MARGINAL TAX RATE Consideration should also be given to reducing the maximum marginal tax rate on all individual income from 70 percent to 50 percent (presently the maximum rate on wage and salary income). That would reduce the maximum rate on capital gains from 28 percent to 20 percent (plus any minimum tax that might apply), and it would permit individual proprietorships or partnerships to operate at much lower tax levels if they elect to develop their inventions themselves rather than sell them to a corporation.

---

[10]/ Charles River Associates, Inc., *An Analysis of Venture Capital Market Imperfections*, prepared for Experimental Technology Incentives Program, U.S. Department of Commerce, February 1976, p. 2.

## SELECTIVE INCENTIVES

A fuller description and technical discussion of some of the following potential changes in selective tax policies is contained in Appendix A.

**CREDIT FOR INVESTMENT IN R & D FACILITIES** Because an investment tax credit is one of the most cost-effective incentives, a permanent increase in the credit from 10 to 20 percent for R & D facilities would stimulate R & D expenditures.*

**FASTER DEPRECIATION FOR PATENTS** The modern pace of technology makes most patents obsolete before the end of their traditional seventeen-year life. We therefore propose that externally acquired patents and intangible technology both be subject to accelerated depreciation over a ten-year period or their demonstrated life, whichever is shorter.

**REQUIRE ONLY DIRECTLY RELATED R & D EXPENSES TO BE DEDUCTED FROM FOREIGN-SOURCE INCOME** In response to the potentially adverse effects of Treasury Regulation Section 1.861-8 on U.S.-based research and development conducted by U.S. multinational companies, only the portion of R & D expenses directly related and traceable to foreign earnings should be treated as a deduction from foreign-source income.

**PROVIDE TAX CREDIT FOR SUPPORT OF UNIVERSITY RESEARCH** In order to stimulate basic research at universities, tax policy could permit corporations to deduct a specific portion of their contribution to nonproprietary research conducted at universities.

**ENCOURAGING PIONEER PLANTS** There is one especially difficult hurdle in the invention and innovation chain. The relatively high cost of a pioneer plant (first-of-a-kind commercial-size facility) poses a special capital formation problem. The costs of bringing new products or processes into commercial-scale production account for an increasingly larger share of the total cumulative cost of inventing, developing, and commercializing a new technology. The combination of the high cost of building the pioneer plant and the high risk of commercial failure in today's inflationary and regulation-prone environment often deters the application of the benefits of research and development.

Special tax incentives could be used to encourage this phase of the innovation process. Such incentives are used in many foreign countries. In Canada, for example, cash grants of up to 50 percent of the total cost of the project are awarded for new or improved products or processes that represent genuine technical advances with commercial prospects. In Japan, investment in new technology is granted a first-year depreciation allowance of up to one-third of the cost of facilities in addition to the normal depreciation allowance. These foreign programs emphasize tax policy that aids commercialization of new technology.

---

*See memorandum by MARK SHEPHERD, JR., page 70.

The importance of pioneer plants in the technological progress of many industries underlines the need to implement a practical, cost-effective way to encourage new-product and new-process inventions. Flexible depreciation and a supplemental investment tax credit would help to bring promising inventions to profitable commercial development.

To date, the questions of definition, delineation, and implementation have plagued suggestions to assist pioneer-plant construction, whether by selective tax incentives or by other means. However, other nations have overcome these obstacles within their existing political and economic institutions, and efforts should be made to find a market-oriented solution.

ENCOURAGING INVENTIVE ACTIVITY IN SMALL BUSINESSES In order to provide incentives that will especially benefit small businesses and encourage them to engage in inventive activity, the following policy changes should be considered:

Increasing the deductibility of capital losses against ordinary income for all taxpayers would assist investors in enterprises seeking radically new processes or products.

The carry-back provisions of the tax code for both net operating losses and investment tax credits could be extended from the present three years to seven years.

Small corporations (Subchapter S in the tax code) should be permitted to raise capital from 100 investors rather than the present 15 maximum.

Like the more general tax changes we have outlined, a number of the suggested selective tax incentives would affect interests other than those of research, development, and innovation. In the past, those other interests such as increasing consumption expenditures have often dominated the determination of the provisions of the tax code. **However, we believe that the increasingly poor performance of U.S. innovation in the last decade is a warning that such a balancing of tax considerations has tipped too far against invention and innovation. We believe the Administration and Congress should correct this imbalance when reviewing tax policy.**

Tax changes in the direction we propose would stimulate economic activity and increase tax revenues in future years. Consequently, they will produce additional taxable profit and income which will offset some of the revenue loss from the original reduction in tax rates. Such "feedback effects" play a critical role in computations of revenue losses, and intelligent tax policy decisions cannot be made without them. **In assessing the feasibility of any tax policy change, the Committee recommends that the feedback effects of tax changes be incorporated in the calculations used by government to estimate the revenue impact of tax proposals over time.**\*

---

\*See memorandum by ROY L. ASH, page 73.

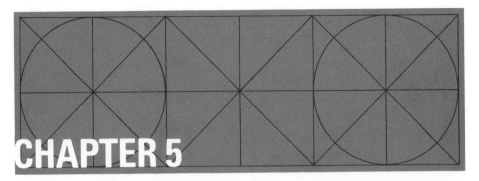

# CHAPTER 5
# GOVERNMENT CONSTRAINTS ON INNOVATION

Many government programs actually constrain innovative activities. Federal, state, and local governments employ an enormous array of controls on various aspects of economic activity. Certainly, some of these regulations serve important economic and social purposes and are needed. Even so, they are frequently inordinately burdensome, and they almost always impose excessive costs on business. Others may no longer serve useful purposes.

Controls impose many kinds of costs on the economy, all of which are ultimately passed on to the consumer. Such costs include, for example, compliance costs and limitations on products and production processes, both of which may make products noncompetitive in international trade, and delays and rigidities in business decision making.

A complete analysis of government economic controls is, of course, far beyond the scope of this policy statement. Therefore, we shall concentrate on the effects of controls on innovation. Some controls are explicitly intended to delay or prevent innovations; others retard innovation as a by-product of their intended effects.

## EFFECTS OF EXCESSIVE GOVERNMENT REGULATION

Traditional economic regulation of such industries as transportation, communications, utilities, and financial institutions largely predates World War II. This older form of regulation is frequently industry-specific, and although it does impose costs on business, it is designed to deal with the economic environment in a particular industry even though it may have consequences in other sectors of the economy. Regulatory agencies are required to consider the future economic performance of the industry as they implement regulatory policy.

In contrast, regulatory policies established in recent years tend to cut across many or all industries and to be for such purposes as environmental and consumer protection and occupational health and safety. Some government regulation of economic activity is necessary. For example, without government-imposed constraints, air and water quality would be significantly lower than society wants.

The cost of achieving health and safety and environmental goals is substantial and is especially high for manufacturing industries. Regulatory agencies are rarely required to consider the future economic success of the industry as these newer forms of regulation are implemented.[1] The cost of these regulations is also ultimately borne by the consumer in the form of higher prices. The cost in terms of slower growth in innovation is less obvious, but it is no less real. CED has recently issued a comprehensive policy statement on the effect of regulation on the market economy.[2]

In some cases, a regulation is specifically designed to stimulate research, development, and innovation. For example, an important goal of environmental legislation has been to induce automobile manufacturers to undertake large-scale research programs on emission control devices and better mileage. But quite apart from the desirability of such induced programs, they add to the cost of the resulting products and divert resources from other research that could lead to greater economic and social advances. Therefore, laws and regulations should not require more research than is justified by society's need for the resulting products of the new technology.

The growth of regulation and the way regulatory policies are implemented inhibit those expenditures in research and development and new plant and equipment that lead to greater productivity through innovation. This disincentive occurs in several ways.

**REGULATION SHIFTS THE ALLOCATION OF RESOURCES**[3] Busi-

---

[1] See George C. Eads, "Chemicals as a Regulated Industry: Implications for Research and Product Development" (Paper presented to the American Chemical Society Symposium on Innovation and the Chemical Industry, Miami Beach, Florida, September 14, 1978).

[2] *Redefining Government's Role in the Market System.*

[3] See Eads, "Chemicals as a Regulated Industry: Implications for Research and Product Development," p. 22.

ness must allocate considerable resources to compliance with regulatory programs. Consequently, funds and skilled personnel are diverted from the testing of applied research product to techniques for meeting mandated standards. Furthermore, some of the newer forms of regulation also specify the type of technology to be used in production. As a result, resources must be used to retrofit existing plant and equipment to meet the best-available-technology standard. Also, investment in plant and equipment that embodies new technology is directed toward expenditures necessary for complying with regulations, and these expenditures may not create the technology that produces the greatest productivity gains.

**REGULATION INCREASES WAITING TIME FOR RETURN ON INVESTMENT** The Food and Drug Administration (FDA) is legally mandated to apply stringent requirements to the marketing of new drugs. Drug innovators are required to demonstrate the safety and efficacy of drugs before they can be marketed. It is, of course, important that the government monitor this crucial activity, and the public needs protection from the unintended consequences of innovations in drugs. But in recent years, FDA policy has prevented drugs from being introduced if they might be harmful in *some* degree to *some* people under *some* circumstances. Too little weight has been given to the potential health benefits that have been forgone because of the failure to make prompt and appropriate use of new drugs.[4]

In 1962, the average R & D cost for each new chemical entity approved was about $4 million. Today, the average cost is $50 million. Moreover, fifteen years ago, it took about two years to bring a new drug to market. Today, it takes seven to ten years.

Extending the length of time needed to recapture the investment in a successful innovation reduces the real rate of the return and therefore discourages business expenditures to increase technological progress. In the pharmaceutical industry, for example, the rate of increase in new chemical entities has declined substantially over the past fifteen years. Furthermore, the United States is lagging significantly behind the United Kingdom and other advanced industrial nations in the introduction of new drugs.

**MANY REGULATIONS INCREASE UNCERTAINTY AND REDUCE INCENTIVES TO INNOVATE** Mandated standards of how an economic activity should be carried out impose obvious constraints on the potential innovator. Regulatory officials yield a very high degree of power in the innovation process because their decisions affect whether a new product or new

---

[4] See *Redefining Government's Role in the Market System*, pp. 68-71 for a detailed discussion of the impact of regulation on innovation in the pharmaceutical industry.

production technique can be commercialized. For example, the law requires the Environmental Protection Agency (EPA) to impose stringent discharge standards on producers. In the absence of clear legislative guidelines on how much of a particular substance should be discharged into the environment and on how the permitted discharge should be allocated among dischargers, EPA is inevitably driven to approving certain devices, configurations of production, and products that will result in small discharge quantities. The result is that plant and product design and the details of production and materials use are subject to government approval. In addition, producers reason that previously approved products and production processes are likely to be approved again and that new products and processes may be disapproved or that their approval may be subject to significant delay. Thus, flexibility and innovation are discouraged by uncertainty.

The uncertainty and instability of regulatory standards are perhaps the greatest disincentive to innovation. Regulatory policy in occupational safety and health, the environment, and consumer protection tends to pursue zero-risk goals. And in their effort to eliminate all risks, regulatory agencies often change standards without regard to cost.

In the case of the acrylonitrile bottle (a type of plastic bottle used for soft drinks) for example, the FDA changed its standards three times between 1976 and 1978. It lowered the migration level for acrylonitrile from 300 parts per billion in 1976, and by 1978, despite evidence to the contrary, it claimed that there is no safe level of migration. Such frequent changes in regulatory standards create great uncertainty that makes business extremely cautious about investing in innovations.

## REGULATORY REFORMS TO STIMULATE INNOVATION

Several reforms are needed to reduce the inhibiting effect of regulation on research, development, and innovation.

The first question to ask about every regulatory program is whether the activity needs to be regulated at all. For example, a consensus has developed that most economic regulation of the airline industry is unneeded, and the Civil Aeronautics Board is being gradually phased out. It is hard to imagine that Interstate Commerce Commission regulation of other domestic transportation, principally road and rail, is more justified. Trucking is inherently a highly competitive industry and is characterized by small producers; most railroads, far from posing a threat of monopoly power, are dependent on government support for their existence.

Where it is concluded that regulation is necessary, such regulation should be carried out in a way that least distorts the market system. In some areas, this means substituting simple yet powerful economic incentives for detailed rules. In the case of environmental programs, for example, the substitution of discharge fees for detailed rules has been studied in depth and recommended by CED.[5] Similar reforms could be devised in other regulated activities, provided, of course, there are minimum acceptable standards of health and safety below which no firm would be permitted to operate. They would accomplish the justifiable social purposes of regulation with less inhibition of innovation and market flexibility.

In addition to considering whether the regulation achieves its primary purposes, **Congress and the regulatory agencies should analyze each proposed regulation for its effect on innovation. Such an analysis should take into consideration the probable effects on investment resources through restrictions on business management's ability and incentive to invest in research, development, and innovation. Only if the social benefits (including the avoidance of harmful side affects) of a regulation will outweigh its full costs should it be undertaken.**

Existing regulatory programs should be examined periodically. Such programs should be modified or phased out if their anticipated benefits fall below their probable costs, including costs of delayed or forgone innovation.

## INTERNATIONAL TECHNOLOGY TRANSFER

There has been much concern over international technology transfer because of a growing fear that it entails the export of jobs. In the 1950s and 1960s, the United States was an international technological leader; today America's lead has shrunk or disappeared in many industries. As a result, U.S. exports have fallen, and imports have increased. There is concern that jobs which should have been created in the United States have instead been created abroad and that U.S. workers have been left unemployed or in less well paid employment. Consequently, government is frequently asked to restrict the export of U.S. technology.*

Technology is transferred among countries through many business and nonbusiness mechanisms. The nonbusiness mechanisms include patent literature, which is as available to foreigners as it is to Americans; university work, such as the training of foreign students; private publications, such as scientific papers, textbooks, and technical documents; conferences, sym-

---

[5]/ See *More Effective Programs for a Cleaner Environment* (1974) and *Redefining Government's Role in the Market System*, Chapter 6, pp. 102–103.

---

*See memorandum by MARK SHEPHERD, JR., page 73.

posia, and lectures, for which Americans go abroad and foreigners come here; career development programs, under which foreign scientists are brought to work on public or quasi-public research projects; government publications, which contain a wealth of detail regarding new technology; and reverse engineering, an increasingly skilled activity in which products are disassembled in order to determine how they are made.

Important business mechanisms by which technology is transferred include product exports, which frequently involve agreements to provide technical documents and to train foreign personnel in the operation and maintenance of the products; trade shows and exhibits, which provide technical information to participants whether the shows are held here or abroad; licensing, under which foreigners pay for technology, including documents, technical assistance, training, updating, use and maintenance technology, and provision of equipment; direct investment by U.S. firms abroad, which usually entails training foreigners to work on U.S. equipment; joint ventures between U.S. and foreign companies; and turn-key projects, in which a U.S. company designs and constructs a project in a host country and then transfers it to a company there.

With such an array of mechanisms, it is clear that government or business restrictions on the transfer of technology would indeed be very difficult to introduce and enforce. This is a free and open society in which people, products, and ideas move without hindrance. A comprehensive government program to control international technology movements would pose serious legal, administrative, and economic problems. Furthermore, the United States now has little civilian technology that is not available in other countries.

Furthermore, pervasive controls on international technology transfer would be harmful to U.S. employment and living standards. CED has long emphasized that international trade and investment are beneficial to the U.S. economy. They enable this country to devote its productive resources to those activities at which it is most efficient, exporting domestic products to buy things abroad that other countries produce most efficiently. Consequently, U.S. incomes and living standards are higher than they could be if international trade was restricted.

Technology transfer is an important part of this beneficial system of international exchange, and Americans have benefited from the flow of innovations to the United States. Business export of technology is an important component of the U.S. balance of payments, and Americans earn at least $4 billion worth of foreign currency a year in payments for royalties and fees for intangible property, much of which is related to technology transfer. The balance of payments for such services is favorable and large. Furthermore, technology transfer complements the export of commodities. Much technology transfer takes place as part of foreign investment by U.S. firms. About one-third of that investment is in finding and extracting natural resources abroad. Another substantial portion of technology transfer takes place in the process of selling complex products abroad.

Technology transfer certainly helps create high-paying jobs in the receiving countries, but not at the expense of jobs in the United States. In some cases, foreign investment in the United States has created jobs for Americans. International trade in general and trade in technology in particular benefit both exporting and receiving countries. More specifically, U.S. firms that invest abroad also benefit by the growth of domestic employment. Some products are not exportable unless the technology used to produce them is also exported. And without the export of such technology, U.S. jobs would be lost. Equally important, technology transfer takes place in a competitive but interdependent world. In many cases, if the technology were unavailable from a U.S. firm, it could be bought from a foreign firm, and the benefits of exporting technology would thus be realized elsewhere. The purchase of new technology from a foreign source can create jobs abroad just as well as purchase from a U.S. source can. Thus, restrictions on technology transfer would mean that jobs would be created abroad and not in the United States.

Many jobs created abroad by foreign investment and related technology transfers could not be created in the United States by simply restricting investment and technology transfer. For example, jobs related to foreign investment in the extraction of scarce national resources can be created only where the resources are found. Restrictions on such investments and related technology transfer would merely mean that the benefits from such foreign investment would go to foreign countries. Similarly, many jobs created abroad by technology transfer are in maintenance, service, and support activities that are tied to the locations of foreign customers. Such jobs could not be created in the United States.

## NATIONAL SECURITY: A SPECIAL CASE

National security is the one clear justification for controls on technology transfer. It is certainly justifiable for the government to prevent weapons system technology from falling into the hands of potential adversaries. Government controls on technology transfer for national security reasons seem to have worked reasonably well and do not need fundamental change. However, elements of unpredictability, delay, and indecision have crept into the system of controls in recent years.

The reasons for national security controls on technology transfer and the limits such controls establish must be as clear as possible in several respects:

- The controlling agency should have clearly stated objectives and should weigh both the benefits and the costs of each proposed control.
- When control of the export of a particular technology is proposed, a precise and detailed specification of that technology should be provided.
- Hearings should be held at which all affected parties can testify on the implications of the proposed controls. Without such hearings, controls may harm civilian users of related technologies. For example, poorly drafted controls on semiconductor technology might curtail exports of sewing machines without protecting any technology important to national security.
- There should be sunset provisions mandating frequent review of control procedures and categories. Because of the dynamic nature of technology, the duration of controls for national security purposes should be carefully limited. Once the national security justification for control has ended, it may be valuable to permit technology transfer for civilian purposes.

International conventions are needed to protect ownership rights internationally, just as patent and copyright laws protect them domestically. Such conventions are also needed to provide incentives for international transfer of technology, just as ownership rights are necessary to provide incentives for domestic technological change. These conventions are controversial, especially between developed and developing nations. The U.S. government should press strongly for maintenance of, and adherence to, international conventions. It has a responsibility to inform industry about foreign national and regional trends or actions that may weaken conventions. Moreover, because abrogation and violation of conventions by other countries weaken the environment of a free international economic system, the U.S. government should take steps to prevent such actions.

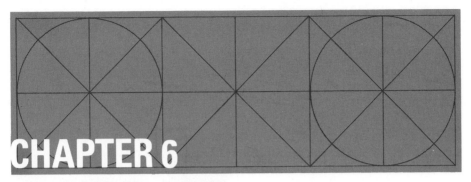

# CHAPTER 6
# PATENTS AND THE INNOVATIVE PROCESS

Patent law is one of the three branches of law that regulate the ownership and use of intellectual property. Trade secret law and copyright law are the two other branches. In the field of technology, patent and trade secret laws overshadow copyright laws.

Intellectual work can be owned and can otherwise assume the elements of property only to the extent provided by law. This characteristic of property, the "appropriability of exclusive rights" to innovative work, largely determines whether such work is reproduced on a commercial scale. If the results of innovative work were freely available to all who might copy or steal them, there would be little reason to invest in research and development. The property rights created under the patent and trade secret laws are thus important, perhaps often essential, to the willingness of prudent businessmen to sponsor research and development and to invest in the facilities needed for innovative products.[1]

There are a number of key distinctions between patents and trade secrets. A patent owner has rights superior to those of a second inventor who makes the same invention; whereas a trade secret proprietor has no rights

---

[1] Although patents generally play an important role in stimulating technological progress, their importance varies among different types of technology. Because of the nature of some technology, the patent may not provide an effective property right.

against a later discoverer of the same trade secret. U.S. patents run for a term of seventeen years from the date of the grant; whereas the life of a trade secret is indeterminate. Patents are obtainable only through disclosing to the public the invention and the preferred manner in which it is practiced; whereas trade secrets depend for their existence on being known to only a few. Thus, if an invention is ascertainable from a product as sold, the only real protection available is through the patent system.

An efficient patent system meets three criteria: The first is accessibility; the system should be simple, inexpensive, and available to everyone. The second is reliability; the system should allow a person who receives a patent, as well as those who are asked to respect the patent, to know the bounds of the protection and to rely on the patent's ability to stand up under litigation. The third is selectivity; the system should protect and encourage significant discoveries without burdening the public with patents on minor and obvious variants of what was previously known. The inventive contribution must be worthy of the protection provided by the government.

A patent is sometimes regarded as a limited monopoly, but in fact, that is rarely the case. Usually, a patent covers only a specific product, product feature, or process in such a way that unpatented design alternatives almost always exist.

## ROLE OF PATENTS

Patent laws (and trade secret laws, when applicable) make developed technology a controllable property. The ability to control a new technology as property determines in large measure whether new-technology development will be undertaken at all.

Patents reward the innovator in a number of different ways. One use of patents is licensing, under which others are permitted to practice the invention in return for royalty payments. Patents are also used to reserve a particular product feature or process for the exclusive use of the patent owner, although probably not so often as is commonly believed. Occasionally, patents are used as trading stock for freedom to operate under the corresponding patents of others.

Another common role of patents is to provide the basis for acquiring counterpart patents in other countries for use in connection with product sales in those countries. Where export sales may be impractical, patents can be used to support licensing in foreign countries.

The number of U.S. patents issued annually has reached a plateau of approximately 75,000. However, the patents issued to foreign corporations have been increasing for some time by more than 1,000 patents a year and are nearing an annual level of 30,000 patents. In 1977, 37 percent of all

U.S. patents issued were granted to foreigners. There has been a steady and noticeable decline in U.S. patents issued to U.S. residents.

There is an obvious correlation between patents granted and research and development undertaken. The recent decline in U.S. patents issued to U.S. residents may be due to the fact that although total R & D activity has not decreased, work in potentially patentable areas has fallen off. Diversions of technology to such areas as environmental protection, energy conservation, compliance with certain government requirements, and development of computer software have had an adverse effect on the output of new patents. Diversion of technology to short-range needs from longer-range, more creative work has also been cited as a cause of the decline.

## CHANGES IN PATENT POLICY TO ENHANCE THE INNOVATIVE CLIMATE

There are several possible ways in which the contribution that the patent system makes to invention and innovation processes can be enhanced.

**FIRST-TO-FILE PATENT SYSTEM** When two or more applicants seek a patent for substantially the same invention, the U.S. patent system provides for interference, a procedure to determine who first made the invention. That party will be entitled to the patent to the exclusion of those who invented later. The interference starts out as a quasi-judicial proceeding in the Patent and Trademark Office (PTO) and occasionally reaches the federal courts as full-scale litigation.

Patent interferences are highly technical proceedings and are of questionable efficacy in determining the first inventor. Much time is spent proving what happened before the filing dates, trying to prove that the inventor was incorrectly named, did not really have the invention in hand, failed to discharge various obligations, and so on. In a significant majority of interferences, the patent is eventually awarded to the first to file.[2]

The United States and Canada are unique among all the industrial countries of the world in utilizing the interference approach. European countries have always considered that a patent should go to the first party to file an application. The new European patent system, which all European Economic Community countries have now adopted, provides a personal defense to the individual who can show he was actually the first to invent and took steps toward use. (Participants in this system include the United Kingdom, France, West Germany, and Holland.)

---

[2] According to a survey by a major corporation, although approximately 110,000 U.S. patent applications are filed each year, only 75 to 80 interference procedures produce a result different from the first-to-file system.

Because the purpose of the patent system is to encourage disclosure to the public, the party who is first to file should be rewarded. Adoption of a first-to-file system would eliminate patent interference proceedings, simplify patent litigation, make patent validity more certain, and serve the interests of the inventor and the public in a more efficient manner.[3*] Most objections to the system could be answered by provision to grant a prior inventor a personal right to use the invention. Such a right would be contingent on not having abandoned the invention and should require proof of steps taken toward commercialization.

**REEXAMINATION OF PATENTS** When considering patent applications, the PTO frequently does not take into account all relevant prior patents or other background material. In the relatively short time that patent examiners have to search for pertinent information, they cannot be expected to find all the prior references, particularly those from foreign countries, that potential defendants in infringement suits may be able to uncover. At present, a patent owner faced with new references brought forward by an infringer can ask for a reissue of the patent to overcome those references. The reissue patent is a substitute for the original patent and expires on the same date as the original, but the scope of its coverage is usually different.

This system should be continued and encouraged. Anyone opposed to the reissue, either wholly or in part, has the opportunity to submit references and to present written arguments. That practice, too, should be continued. We do not suggest that a full adversary proceeding be established in the PTO; such a system would increase both the time and expense involved in the reissue proceedings.

If the patent for which reissue is sought is already in the courts, the judge should retain the discretion to stay the action and allow for the reissue. Such a procedure may often allow for greater certainty in the proper coverage of the patent and consequently reduce the amount of judicial time required. On other occasions, the reissue procedure may be insufficient because oral testimony, for example, may be essential to explain the references and other background matter submitted by the defendant. In such cases, the judge might properly refuse to stay the suit. **We believe that the present procedure is reasonably satisfactory insofar as the patent owner is concerned and that it should not be changed with respect to their opportunities.**

---

[3]/ The pharmaceutical and agricultural chemical industries have questioned the applicability of the first-to-file system. They believe that the present system is preferable in their businesses, given the extensive testing required to prove practical results.

---

*See memoranda by J. PAUL LYET and MARK SHEPHERD, JR., page 73.

However, a new right should be provided to defendants or potential defendants faced with adverse patent claims. **If defendants feel that their references are strong enough to invalidate the patent without the need for, or expense of, a trial, we recommend he should have the opportunity to take his references to the PTO and ask for reexamination of the patent in light of those references.** In other words, the defendant should be able to call for a PTO action (akin to the reissue proceedings) that might result in a PTO decision that the patent should not have been granted. In certain situations, particularly when patents without merit are involved, such proceedings would save the time and expense of a trial. To that end, the judge should have the discretion to stay any trial proceeding pending the outcome of a PTO reexamination requested by the defendant. The reexamination procedure should require that the person requesting it pay fees that will approximate the PTO costs involved.

In recent years, the concept of a reexamination of issued patents appeared in various congressional bills, some of which have been supported by the American Patent Law Association; the Patent, Copyright and Trademark Section of the American Bar Association; the New York Patent Law Association; and others. **We support the efforts of Congress to improve this feature of patent policy[4] and believe that a reexamination procedure will increase certainty in the patent system at a relatively modest cost to all involved.**

ARBITRATION OF PATENT DISPUTES Commercially important patents often invite controversy. Competitors interested in the patented product or its equivalents frequently disagree with the patent owner concerning its true scope and value. A suit in the federal courts may result. Protracted pretrial procedures and the difficulties of dealing with technical subjects tend to make the litigation very expensive, often $500,000 or more for each party.

Patent litigation itself is also protracted. A suit on an important patent will commonly take years to resolve. The cost and time required for litigation detract from the ability of patents to foster innovation. The cost of enforcing patents may very well influence business to invest in programs that are less risky than the research and development needed for innovation and productivity gains.

---

[4]/ Through proposed rules published as 43 Fed. Reg. 59401 (December 20, 1978) and later withdrawn, the Commissioner of Patents and Trademarks raised the possibility that a limited reexamination procedure could be established in the PTO based on his rule-making authority. However, we believe a statute is needed because the commissioner cannot provide for court review of PTO decisions or authorize cancellation of patents found to be invalid, features that are required for an acceptable reexamination system.

The patent owner shoulders the burden of detecting infringement and enforcing the patent through civil action in a federal district court. The federal courts have not been inclined to share their exclusive jurisdiction over patents because the validity of a patent appears to hold such public interest that settlement of a controversy through arbitration is inappropriate. The cost of enforcing or defending against a patent in court could amount to several hundred thousand dollars for each party. To reduce the cost and time required for the resolution of certain patent disputes, arbitration should be available for those who wish to use it. **We believe that arbitration should be a legitimate method for solving patent problems.** Arbitration is common in resolving disputes in almost all other commercial areas, including very large labor settlements, and it is difficult to see why it should not be allowed for patent questions.

Although compulsory arbitration cannot be required because such a requirement would be a violation of due process, it should be available when both parties wish to use it voluntarily. The results of arbitration are, of course, binding only on the participants. **We therefore recommend that public policy be modified to permit voluntary arbitration of patent disputes, including questions of both infringement and validity.**[5]

## ADDITIONAL RECOMMENDATIONS ON PATENT POLICY

This Committee believes that a number of additional changes in patent policy would contribute to greater innovation in the U.S. economy. Many of these changes may appear quite minor in themselves, but their combined effects on future innovation could be quite substantial. These recommendations and the analyses supporting them are discussed in Appendix B "Supporting Analyses for Additional Changes in Patent Policy."

In summary, this Committee recommends the following additional patent policy changes:

- A single patent appeals court should be established to eliminate the current problem of inconsistency in precedents between existing Circuit courts.

- The recommendations of the 1978 National Commission on New Technological Uses of Copyrighted Works should be implemented to provide legal protection to the authors of computer programs and to

---

[5]/ To our knowledge, very little data on arbitration in patent cases exist. However, information provided by the American Arbitration Association shows that in a sample of 200 commercial arbitrations in the construction field, it averaged less than four months from filing to issuance of an arbitrator's award. The savings in time and money would be substantial if anything like this could be attained in patent cases.

assure that rightful processors of copies of computer programs can use or adopt those programs for their own use.

- U.S. owners of patented processes should be able to enforce their patents against goods made abroad with those processes and then imported into the United States.
- A patent owner should be able to receive an extension of a patent's life equal to the length of government regulatory delays.
- So that government-funded research and development will be used for commercial products, government contractors should in most instances receive title to the inventions and patents made under government contract.
- Government-owned patents should be licensed to all domestic manufacturers on a royalty-free basis.*

The Committee also feels that given the PTO's importance in the innovation process, it is vital to provide adequate funding to eliminate the unnecessary backlog in the work of that office.

Proposals for *mandatory licensing* of government patents and the *reimbursement of government seed money* when commercialization of patents that received government assistance is successful do not appear to us to have any merit. We also reject the concept of *periodic maintenance fees during the life of patents* as a means of reducing the number of worthless patents that continue on the rolls since there would be no positive effect on domestic industrial innovation and foreign competitors might be assisted in U.S. markets.

There are two features of patent policy that we believe deserve examination in future policy development. First, the concept of *shorter-term lesser patents,* which is used in Germany and Japan, should be considered for future introduction in the United States. Such patents are designed to protect minor advances in technology and are worth evaluating. Second, although the Committee offers no specific policy recommendations on how to improve *patent protecton in Third World countries,* we do believe that the government has a responsibility to work with other industrial countries in dealing with this potentially important issue. We believe that the United States should encourage, where appropriate, the establishment of effective laws to protect property rights in inventions and innovation while avoiding any improper interference in the affairs of other countries. This should include the establishment of effective patent laws in Third World countries.

---

*See memorandum by JOHN D. GRAY (Omark Industries), page 74.

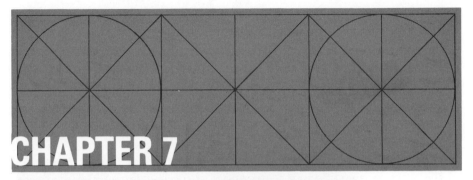

# CHAPTER 7
# DIRECT GOVERNMENT SUPPORT OF RESEARCH AND DEVELOPMENT

The federal government finances about half of all research and development in this country. In fiscal 1979, federal support amounted to about $25 billion, or more than 1 percent of GNP.[1] Government support for basic research is justified because much of it eventually does produce important social benefits. However, the nature of those benefits is almost impossible to foresee. Almost all countries devote some public resources to research whose only justification is the improved knowledge it provides about the universe.

Such motivations have been enough to spawn an enormous variety of government R & D activity. To assure maximum efficiency from that activity, we believe certain guiding principles need to be applied (see the section "Guiding Principles for Government Support of Research and Development" later in this chapter).

## TRENDS IN GOVERNMENT R & D SPONSORSHIP

Federal R & D spending as a share of GNP has fallen by about one-third since the mid-1960s (see Chapter 2). But as the data in Figure 10 make clear,

---

[1] State and local governments also provide substantial amounts of research funds, particularly through state universities. Statistics cited here do not include state and local activities because the relevant data are not available in sufficiently meaningful detail. However, most of the principles and recommendations set forth in this chapter can be usefully applied at all three levels of government.

**FIGURE 10**

**Allocation of Federal R & D Expenditures, by Individual Function, 1969 and 1978**

| (percent) | 1969 | 1978 |
|---|---|---|
| National defense | 53.4 | 49.0 |
| Space | 23.9 | 11.9 |
| Energy | 2.1 | 10.6 |
| Health | 7.2 | 10.2 |
| Environment | 2.0 | 4.2 |
| Science and technology base | 3.3 | 4.0 |
| Transportation and communication | 2.9 | 3.1 |
| Natural resources | 1.3 | 2.3 |
| Agriculture | 1.4 | 1.9 |
| Education | 1.0 | 1.0 |
| Income security and social services | 0.6 | 0.6 |
| Community development, housing, and public services | 0.3 | 0.4 |
| Economic growth and productivity | 0.4 | 0.4 |
| International cooperation and development | 0.2 | 0.3 |
| Crime | — | 0.2 |
| Total* | 100.0 | 100.0 |
| Total spending (millions of current dollars) | $15,641 | $26,317 |

*Due to rounding totals may not equal 100.

SOURCE: National Science Foundation, "An Analysis of Federal R&D Funding by Function: Fiscal Years 1969–1979."

the recent decrease in federal R & D spending has been strongly concentrated in space and to a lesser extent in defense. All other categories have either increased or remained constant as a share of total federal R & D spending and have almost kept pace with the rise in GNP.

A shift in the basic nature of federal research spending has accompanied the shift in functional mix. As the data in Figure 11 show, the percent of federal support that goes to basic and applied research has risen somewhat since the mid-1960s and the percent going to development work has fallen.

The slight increase in the proportion of federal R & D spending devoted to basic research is consistent with the recommendations made by this Committee. However, we believe there is reason to be concerned about the decline in R & D spending as a proportion of GNP.

Although it is impossible to specify the extent to which defense and space expenditures for R & D and procurement of innovative technologies have contributed to the rate of technological progress in the United States, there would seem to be reasonable support for the view that progress made in fields such as electronics and computers was accelerated by the high levels of federal effort in the 1950s and 1960s.[2]* Because government re-

---

[2] A number of studies have documented the economic returns from government R & D in space exploration. Advances from research in specific areas such as cryogenic insulation, gas turbine engines, integrated circuits and computer programs for structural analysis are documented in *Quantifying the Benefits to the National Economy from Secondary Applications of NASA Technology* (Mathmatica, Inc., March 1976).

---

**FIGURE 11**

**Allocation of Federal Expenditures by Category of R & D, 1965, 1970, and 1978**

| (percent) | 1965 | 1970 | 1978 |
|---|---|---|---|
| Basic research | 13.9 | 16.7 | 17.0 |
| Applied research | 19.4 | 21.0 | 23.8 |
| Development | 66.7 | 62.3 | 59.2 |
| Total | 100.0 | 100.0 | 100.0 |

SOURCES: National Science Foundation, *Science Indicators 1976* and "An Analysis of Federal R&D Funding by Function: 1969–1979."

---

*See memorandum by J. PAUL LYET, page 74.

ductions of the late 1950s and early 1970s have not been offset by increases in the private sector's support, we assume that some social and economic benefits have been, and will, be forgone.

The role of procurement of advanced technology products by these agencies cannot be overlooked, however, and we do not wish to imply that replacement of the R & D support alone, without regard to the mission needs of the government, would be an efficient use of public funds. Rather, we would suggest that appropriate incentives for private technological innovation which would rely on the market system for the allocation of resources should be considered to counterbalance such shifts in federal policy. Because the long-term nature of the payoff from R & D tends to obscure the impact of major reductions in the nation's technical efforts, greater attention to coordination of policies affecting technological progress in the economy appears to be needed.

## GUIDING PRINCIPLES FOR GOVERNMENT SUPPORT OF RESEARCH AND DEVELOPMENT

Research and development to improve products and production processes for market activities must be the responsibility of industry. Industry is in the best position to make R & D decisions that must be coordinated with production and sales activities because it can judge better than government which activities are likely to lead to products and processes that can be manufactured and sold. In general, therefore, the role of government with respect to research and development aimed at commercial applications ought to be restricted to broad-based indirect incentives designed to take advantage of the inherent efficiencies of the market system (see Chapter 2).

However, this principle still leaves several areas in which the government can properly play a more direct role in financing research and development.

**DIRECT FEDERAL NEEDS** The government does have a role to play in developing those innovations that are primarily for government's own use. Defense and space research and development are the most noteworthy examples, but research in such activities as mail delivery, air traffic control, regulatory standards, and sewage disposal also falls in this category. In these areas, government is the purchaser of products and processes that result from the new technology and is thus in the best position to evaluate R & D needs for the same reason that business is in the best position to evaluate R & D needs in areas of commercial application. In both cases, decisions should be in the hands of those who are responsible for the products and processes that will be developed or improved by the R & D activities. That means lodging chief responsibility in the hands of the particular agencies and holding them accountable as part of their regular budgetary and performance reviews.

**ENSURING SOCIAL BENEFITS FROM BASIC RESEARCH** A second area in which government must play an active part in financing is basic research. By its nature, fundamental research involves highly uncertain results and a high risk of failure. As we have noted, much successful research of this type has eventual government or commercial applications, but usually only after much additional applied research, development, and capital investment over long periods of time. Although a good deal of basic research is privately financed, the connection with application is, in many cases, so remote and ownership rights are so tenuous that the financing of costly basic research cannot be economically justified by the private sector. In addition, there is a strong and intimate connection between the conduct of basic research and the education of scientists and engineers. Such talent, with the ideas it embodies, is a major national resource; we believe, therefore, that it is justifiable for the country at large to pay some portion of the cost of developing that talent through the financing of basic research work in the universities.

The application of these principles to research support for the basic academic disciplines is best left primarily to the National Science Foundation and other federal agencies that finance basic research. The widely used system of research choice based on the review of proposals by fellow scholars has generally served the country well. This system of resource allocation can perform its function even better if agencies are held accountable in budgetary reviews for the broad nature of the supported research, rather than for individual projects chosen.

**SPECIFIC-PRIORITY NATIONAL NEEDS** There is a gray area in which industry research and development would be inadequate even with appropriate broad government incentives. Health, environmental protection, earthquake prediction, and agriculture are some areas in which the government has made this judgment. A broad-based tax incentive for private efforts would certainly not motivate the same amount of research and development in all areas. For example, in some industries, such as agriculture, firms may be too small to undertake research and development, even though valuable new technology can be discovered at modest cost. There may be other justifications for government sponsorship of research and development that are specific to other areas. Each case must be examined on its merits, and those scrutinizing such proposals should apply the strictest and most pointed criteria to test whether the probable social benefits will outweigh the costs.[3]

## BASIC RESEARCH AND THE UNIVERSITIES

It is nearly impossible to predict whether a particular basic research project will lead to useful innovation. But it is clear that socially valuable innovations flow from high-quality basic research. Postwar innovations in atomic energy, computers, chemicals, pharmaceuticals, and other areas would have been impossible without a foundation of basic research results.

In recent years, the costs of basic research have been rising rapidly. Many research projects in the physical and biological sciences require large and frequently interdisciplinary research teams, increasingly sophisticated instrumentation, and special facilities. Moreover, government regulations regarding the environment, health and safety, and the use of human and animal test subjects have added to those costs.

Basic research is by no means financed solely by the federal government; industry, private foundations, private university endowments, and state university budgets are also important sources of funds. But as a practical matter, there is no realistic alternative to the federal government as a primary source of money. We urge that the government explicitly state that a consistently high level of support for basic research is a major national objective. We believe that over the long run, consistent, stable financial support will provide the country with a strong base of knowledge upon which future economic growth and industrial innovation can rest.

Much government-supported basic research is concentrated in universities. There are two principle reasons for this. First, universities have the capability and the freedom to carry out basic research without the competitive pressures of bringing products to the marketplace. Second, there exists a highly complementary relationship between the performance of basic research and the education of engineers and scientists. Those who train engineers and scientists must have contact with the frontiers of knowledge that the conduct of basic research provides. In addition, education of graduate students depends on their active involvement in research projects.

**It is therefore concluded that the federal government should increase its relative funding of basic research, especially in universities, even at the**

---

[3] Although criteria must be more or less tailored to each case, responsible officials can and should draw guidance from such efforts to specify criteria as can be found in the Federal Nonnuclear Energy Research and Development Act of 1974; U.S. Office of Technology Assessment, *The Role of Demonstrations in Federal R & D Policy* (Washington, D.C.. July 1978); and *Redefining Government's Role in the Market System.*

**cost of other activities.** Increased funding for scientific and technical instrumentation should be given high priority in the federal budget. There is a need to redress the effects of earlier cutbacks and to offset the rapidly increasing cost of sophisticated scientific instruments. Agencies making federal grants should consider giving preference to proposals for sharing of instruments whenever that will maximize the return on such an investment. Another consequence of earlier cutbacks in funding is that university research facilities badly need renovation. Federal funds should be made available for this purpose, and priority should be given to renovations that decrease operating costs and energy use.

The current practice of government funding of university research through a cost-sharing arrangement often requires the use of nonresearch university resources. This practice is shortsighted and is an ineffective way of achieving research objectives. Federal support of basic research at universities should provide for full recovery of direct and indirect costs of research funded. The cost of obtaining and administering federal basic research funds can be lessened if government oversight of accountability focuses more on the selection of general scientific areas to be funded, on the procedures employed by universities in administering projects, and on the assessment of research results. Detailed control of the conduct of such research projects is counterproductive. There is an obvious need to require universities to account for how public funds are used. However, in the case of complex projects, the government should be more willing to make a long-term commitment to research support.

Greater emphasis should be placed on financing scholars who demonstrate outstanding research capabilities and less on equality among a large group of institutions. Consideration should be given to developing new mechanisms for federal funding that will maintain the valuable features of peer review yet reduce the cost of proposal preparation.

Finally, the government should encourage universities and industries to consider developing cooperative research programs. Such mechanisms, if properly designed and targeted, ought to improve the access of industry to basic research work carried out in universities. (One tax incentive to achieve this goal is discussed in Chapter 4, page 41.)

## FEDERAL SUPPORT FOR LARGE-SCALE APPLIED R & D PROGRAMS

There are frequent proposals for the federal government to support new large-scale applied R & D activities. We recognize the possible existence of specific-priority national needs that can extend beyond the bounds of conventional government activity (e.g., new energy technologies or advanced aircraft technologies). Each such proposal must be analyzed on its own merits. But in analyzing proposals for direct support, it is necessary first to ascertain whether the proposed activity could be adequately performed by the private sector. It must then be determined whether an adequate performance could be achieved by less intrusive government assistance (e.g., the tax incentives recommended in Chapter 4). Finally, it must be asked whether the probable social benefits of the proposed government program would significantly exceed the probable costs.

The answers to these questions may be yes in some cases and no in others. For example, some projects may be too risky for individual firms in the private sector to finance, even with tax incentives.[4]

Many people also believe that some industrial technology needs cannot be met by the private sector alone, even with appropriate tax incentives. Accordingly, they advocate broad-based industrial technology centers jointly financed by government and industry to undertake nonproprietary applied industrial research.[5] We are opposed to proposals for direct federal support that would make technology and its development an end in itself. New technology is a means to higher living standards and an improved quality of life, and the market mechanism generally provides better guidance in these matters than government does. However, we are not prepared to reject the possibility that in some areas of industrial technology, the unmet needs for major advances are so critical that bolder federal support is justified. In such areas, very special attention should be given to the principles and criteria outlined in this chapter.

Present government policy encourages the use of government procurement activities to stimulate innovation. Government is a large buyer of some products that also have civilian markets. To the extent that it can be done without detracting from the need for efficiency in government procurement, purchases in civilian markets may be used to stimulate innovation.

Private sector invention and innovation could also be helped by increased government information programs. For example, publication of international comparisons of technical activity and the collection and evaluation of data to assess technological progress is important to national policy making.

---

[4] See, for example, the policy statement *Helping Insure Our Energy Future: A Program for Developing Synthetic Fuel Plants Now* (1979).

[5] Herbert Holloman et al., *Government Involvement in the Innovation Process: Policy Duplications* (Cambridge, Mass: MIT Center for Policy Alternatives, 1978).

## FEDERAL R & D PERFORMANCE

Often, the distinction between federal support for research and development carried on in government laboratories versus support for such activities carried on in universities, industry, and other laboratories is not clear. Some government research and development is carried on in institutions that are clearly government-owned and -operated; agricultural and health research laboratories are examples. But much federal support (about $1 billion a year) goes to research laboratories administered by universities. Such institutions undertake both nuclear weapons and power research, and that research is conducted by government–private sector institutions. However the relationship between the two sectors is not the same in each case.

As in the case of other government activities, a common concern is whether R & D activities can be decreased, stopped, or transferred once the need has been met. To help assure such flexibility, we suggest two guidelines for decision making:

- Government should maintain enough in-house research capability to be an intelligent consumer and manager of the research it finances.

- Research projects that require capital investments too large for firms in the private sector to justify, such as high-energy accelerators and astronomy centers, need a degree of government control, and perhaps ownership, not required by other projects.

We believe that there should be means of assessing the role of federal R & D laboratories. Although a comprehensive review of all such facilities would be a massive and perhaps impossible task, there should be more periodic evaluations by visiting committees, peer evaluation, publication, and on-site reviews by congressional and executive branch officials. More efforts also need to be made both in government laboratories and in government-financed research at universities to transfer nonsensitive technological advances to the private sector and to state and local governments.

Inflexibility is a special problem with government research facilities. It has proved difficult to reduce staff and activity levels in response to changes in government spending priorities. Too often, needed budget cuts are implemented by reducing support of university research rather than of government research institutions. The historical pattern of continuous support of federal laboratories throughout periods of major funding reductions indicates that resource allocations have frequently been made on the basis of administrative criteria rather than on the merits of the research projects.

A second problem with government research is its quality. Within all research organizations some government laboratories do very high quality research, but others do routine or unimportant research year after year. In the case of government laboratories however the political system finds it difficult to support high-quality work and to phase out low-quality work.

Both difficulties need to be more vigorously addressed as part of the ongoing operation of government laboratories.

## CONCLUSIONS

This Committee's views on direct federal support for research and development can be summed up as follows:

- Direct support of *basic research* is a necessary and appropriate activity of government. This needs to be a clearly stated national policy, and programs to carry out that policy should be given a high priority in the federal budget. Cost increases because of the increasing complexity of basic research and because of federal environmental and other rules warrant more federal support than is currently being provided.

- The government also has a direct role to play in *applied research* and development, but its activities in that area need to be carefully evaluated according to the guiding principles and criteria we have identified. Such activities are most appropriate in the fields in which government itself will be the major purchaser of the products and processes that result from the new technologies (e.g., defense, space, health, air traffic control, mail delivery, sewage disposal). In expanding direct government support beyond this range, caution is needed in order to avoid interference with private sector research and development.

- Improvements can be made in the effectiveness and efficiency of *government-university relationships,* particularly in the awarding and management of research grants. There are also opportunities for increasing cooperation between universities and industrial firms in research.

We see selectively enlarged direct federal support of research and development as an important means of meeting the government's responsibilities and also as a positive contribution to private sector innovation. However, we believe that the other tax, regulations, and patent policy recommendations offered in this statement are even more important in spurring the introduction and diffusion of new technologies. Taken together, these recommended changes in public policies could do a great deal to bring about a new era of U.S. technological progress.

# MEMORANDA OF COMMENT, RESERVATION, OR DISSENT

Page 4, by ROBERT R. NATHAN

Sufficient emphasis has not been given to the pernicious role of inflation in discouraging capital investment. Tax policies, serious energy crises, and other uncertainties are very important but the most crucial deterrent to investment is the persistent high level of inflation and the lack of prospects for progress toward reasonable price stability in the foreseeable future.

There are many serious consequences of an economic, social, and political nature flowing from high rates of inflation. Perhaps its most clearly identifiable negative impact has to do with investment. High interest rates, difficulties in floating equity securities, the tendency of government policies to fight inflation with recessions, the drop in the value of the dollar, all relate to inflation and all serve to discourage new investment.

Of course, this statement is not designed to focus on inflation. Yet, intensive efforts to pursue price stability are less likely to be forth-coming if policies are proposed which accommodate to the existence of inflation. Many justifications expressed herein for the proposed tax change allowing faster depreciation write-offs sound too much like escalation clauses and indexing and other means of living with rather than fighting against inflation. In essence I would prefer greater emphasis on the cancer-like disease of inflation as the major deterrent to investment.

Page 6, by MARK SHEPHERD, JR., with which ROY L. ASH has asked to be associated

While there is no doubt that greater expenditures on plant and equipment accelerate the diffusion of innovation throughout industry, it doesn't necessarily follow that the line of causation runs from new investment to innovation. Often, the reverse is true: technological advances stimulate increased capital formation. Also, the incorporation of advances in technological knowledge into existing equipment and facilities is directly responsible for a large part of our productivity gains.

## Pages 7 and 30, by RODERICK M. HILLS, with which FRAZAR B. WILDE and D.C. SEARLE have asked to be associated

I approve the statement except that many of the proposals for new tax policies set forth in Chapter Four seem to me to ignore the larger problems. Until taxation of negative income (e.g., taxation of capital gains that because of inflation is not real income) is eliminated, until the tax discrimination against equity capital (double taxation of corporate profits paid in dividends) is eliminated, and until we have real tax incentives for saving, we will not have sufficient resources to sustain technological progress no matter how many incentives we give to inventors. I do agree that special tax credits should be provided for "pioneer" plants.

## Page 7, by RALPH LAZARUS, with which ROBERT R. NATHAN has asked to be associated

While I concur with the importance of this measure, I also feel strongly about the importance of having capital recovery for buildings.

In an August 1978 study by Robert R. Nathan Associates, Inc. entitled, 'Economic Impact of Applying Investment Tax Credit to Retail Structures,' the need to apply tax incentives to both structures and plant equipment is summarized. The report states:

> Economists generally agree that there is serious need for stimulating more investment demand and more investment activity in the United States. A higher level of investment is needed not only to expand employment but also to combat inflation by modernizing and expanding productive capacity, as a means of lowering costs in the United States and making our industries competitive with industries abroad, both in American and overseas markets.
>
> Moreover, there is evidence that among the several ways of stimulating investment in plant and equipment, the investment tax credit is the most direct and most efficient. Consequently, it is in the public interest that it should be continued and that its scope should be broadened to apply not only to equipment but also to structures, in the retail sector as well as in manufacturing. This would serve to stimulate an increase in the level of investment on a wide scale, throughout most sectors of the economy and widely dispersed geographically.
>
> An incentive which affects so large and important a segment of the economy as retailing will have a substantial impact on the economy as a whole. Applying the investment tax credit to retail structures will not only add to capacity, employment, and productivity in that sector but

will also reverberate through construction and other industries, including manufacturing because it provides markets for manufactured goods.

In recent years the volume of commercial construction has lagged. Although business investment in equipment has been increasing since 1971 at an annual rate of about 5 percent (exclusive of the effect of price increases) the volume of commercial construction, other than office buildings, actually inclined nearly 30 percent between 1973 and 1977. Clearly, incentives are needed to stimulate investment in structures by making it more profitable.

I concur with these findings and believe that capital cost recovery is an important example of one such tax incentive.

---

### Page 7, by ROBERT R. NATHAN

The recommendation to permit more rapid capital recovery allowances is highly desirable. However, there are two important objectives that need special emphasis. One concerns the desirability of more rapid obsolescence. We desperately need to replace our old and inefficient plants and equipment more speedily if we are going to restore our competitive posture in world trade. Business will be inclined in junk or otherwise dispose of production facilities sooner if they are totally or near fully depreciated. That is desirable.

Secondly, this economy needs more vigorous competition. A step-up in the rate of investment should contribute to added capacity and to more rapid modernization which in turn should strengthen the forces of competition. If we can achieve more price competition we will have a much better chance of success in fighting inflation. The stimulation of investment can and should contribute to that objective. Maybe incentives for new investment will divert attention of business leaders from the merger mania to more constructive investment programs. That, too, would tend to increase competition and make for a more effectively functioning marketplace.

---

### Pages 7 and 41, by MARK SHEPHERD, JR.

The Committee's recommendation to change current tax policy for depreciating R & D structures and equipment should be broadened to include a direct tax credit for R & D spending. R & D is the catalyst that makes step-function productivity gains possible. According to Professor Kendrick of George Washington University, roughly two-thirds of all advances in managerial and technological knowledge (which, in turn, accounted for roughly half of the U.S. increases in productivity between 1948-1978) stem from formal programs that increase the stock of R & D. Texas Instruments' experience has confirmed that money invested in R & D is highly leveraged in its beneficial effect on productivity.

Texas Instruments Incorporated sponsored a study, prepared by Andrew Brimmer in cooperation with Data Resources, Inc., which demonstrates the positive impact of alternative R & D tax credits (10%, 25%, 50%) on R & D industrial expenditures, productivity, and real GNP growth.

For example, the enactment in 1966 of a continuing 25 percent tax credit on industrial R & D expenditures would have added 0.2 percentage points to annual productivity gains during 1966-77, 0.3 percentage points per year in 1978-87, and 0.4 percentage points per year in 1988-97. To put these numbers in perspective we only need to recall that the total productivity increase over the past seven quarters has been close to zero. This 25 percent tax credit would, moreover, generate, for comparable time periods, average annual gains (in 1972 dollars) of about $2 billion, $5 billion, and $11 billion in R&D expenditures; and approximately $4 billion, $36 billion, and $102 billion in GNP.

Finally, we estimated that the tax cost of this program would average a net loss of $2.3 billion annually for the first ten years. In subsequent time periods, however, the cumulative impact of R & D would begin to pay large dividends. Faster economic growth would produce larger tax gains and the net tax impact would become a positive $6.1 billion per year in the second decade, offsetting by roughly a 3-to-1 margin the tax losses in the previous period. These results indicate that the R & D tax credit is an investment that would yield significant positive returns to society as well as to private firms.

---

## Page 10, by D. C. SEARLE

While I generally support the findings and conclusions of this study, and believe that the CED has made a real contribution toward understanding and dealing with some of the microfactors which inhibit innovation, the study fails to sufficiently emphasize the total impact of government regulation and the resulting environment. By its nature, the study was designed to focus on those facets of our economy which are most closely related to continued technological progress. The adverse effect of government's interference with the market system was mentioned by reference to another CED study. Nevertheless, it is unlikely that this country will ever regain its technological leadership so long as we continue to attempt to solve perceived problems by detailed government intervention.

The mass of government regulations contained in tens of thousands of pages of the *Congressional Record*, untold pages of the *Federal Register*, and millions of pages issued by the regulatory agencies themselves, literally saps strength from the productive sector of our economy leaving little energy and less incentive to undertake the risks inherent in developing most technology. This is the issue which must be faced by the American people and their elected representatives.

### Page 18, by HENRY B. SCHACHT

I am in basic agreement with this CED policy statement. But I would add two additional points. I do not think we know enough about which of the five phases of technological progress (page 13) are the laggard ones in our economy and are having the most serious impact on productivity decline. My guess is that there is no single answer but that it differs by industrial sector and product group. While most of the CED tax proposals would spur all five phases, more targeted public policies are probably needed in sectors where we can maintain or obtain, as an economy, a competitive advantage over our foreign competitors. Increasingly, both European, Japanese, and developing country governments have gained sector by sector expertise in their Ministries of Industry or the equivalent.

And the results of such competent analysis are used to focus trade, tax, and other regulatory policies. My guess is that until the U.S. government develops similar competence (for example, in a revitalized and reorganized Department of Commerce) we will try to spur productivity through technology incentives both too indiscriminately and, in areas of opportunity, too meagerly. Hence, I would have added a discussion of this additional public sector capability as a fundamental need in our efforts to understand and spur technological progress.

### Page 33, by ROBERT R. NATHAN

The report fails to discuss the important problems of depreciation policies with respect to purchases of used plant and equipment. Many legislators and other public leaders will oppose accelerated depreciation unless the benefits are confined only to initial investors in the new facilities. Others will want to encourage acquisition of secondhand capital goods. Presumably the speeding up of obsolescence attributable to more rapid recovery allowances could serve to depress the market for used plant and equipment. That could make it easier for purchasers of used equipment to compete with buyers of new equipment and perhaps slow the pace of new investment. This whole subject of depreciation policies for subsequent purchasers needs thoughtful consideration.

### Page 42, by ROY L. ASH

This policy statement contains a number of tax reduction proposals. They are good ones. The statement would be more complete if it quantified the amount of tax revenues that would be lost to the government from these changes, and then took a position to either:
   a. increase specified other taxes by an equal amount;
   b. reduce specified expenditures;
   c. increase the budget deficit; or
   d. make a case that offsetting revenue gains would flow from the changes.

As we all know, government decision makers must deal with the whole issue. Thus, I would encourage continued efforts by those in government and the public policy field to develop an objectively quantified study of the possible foregone tax revenues from these recommendations and the way they best could be dealt with.

### Page 47, by MARK SHEPHERD, JR., with which ROY L. ASH has asked to be associated

I agree that technology transfer should be encouraged as long as it does not interfere with national security. However, it should be stressed that technological know-how is owned by private companies. It is not owned by governments. Thus, it must be transferred only under conditions that recognize and respect the rights of ownership and control by prospective suppliers.

### Page 54, by MARK SHEPHERD, JR.

Interference practice at Texas Instruments Incorporated has been a minor expense. The implementation of a personal right to use the invention which 'would be contingent on not having abandoned the invention and should require proof of steps taken toward commercialization of the invention' would add more than offsetting administrative cost to document inventions in order to establish a personal right.

### Page 54, by J. PAUL LYET, with which D. C. SEARLE has asked to be associated

Although I concur with many of the recommendations contained in this policy statement, I do not believe that adoption of the first-to-file system would result in the advantages claimed for this approach. I therefore feel that changing to first-to-file would not be an improvement over the present system.

---

## Page 57, by JOHN D. GRAY (Omark Industries)

Royalty-free licensing of government-owned patents which require additional development work and commercialization really does not provide incentive for a manufacturer to undertake the pioneering process and then let other competitors, several years later, come in free on his efforts. I suggest a policy on those patents which need pioneering work and commercialization stating that the pioneering licensee be given a royalty-free, five-year lead period with the five-year period to run from the time the product is first put on the market in a reasonable quantity and with the requirement that the licensee must commercialize the concept within two years after being licensed. During that five-year period, the pioneering licensee would have the right to sublicense and keep one-half the royalty and pay the government the other one-half.

At the end of the five-year, royalty-free pioneering license period, the pioneering licensee, his sublicensees and all newcomers would then be licensed at competitive industry rates with the pioneer licensee to receive a 25 per cent lower rate as compensation for his early pioneering commercialization work.

This suggested policy would apply to all patents not deemed important to national defense, security, etc.

---

## Page 60, by J. PAUL LYET

An important additional source of technical innovation in the United States is the Independent Research and Development Program (IR&D) conducted by companies engaged in work for the Department of Defense (DOD). A great many of the country's major companies are so engaged and have very significant IR&D programs. Because this program can contribute so significantly to technical innovation, any improvements in the way it is administered by the DOD can have an immediate effect on the progress of technology in our country. The spin-offs from IR&D activities have considerable value in the entire spectrum of technical work which companies engage in. The experience of the space program bears out this point.

An administrative problem which the IR&D Program encounters is the government's insistence not to pay companies for IR&D costs unless the work has, in the opinion of the Secretary of Defense, a potential relationship to a military function or operation. The CED should recommend the removal of the requirement for potential military relationship in the IR&D Program, letting the competitive marketplace judge the technical quality and relevance of the program. This will encourage broader gauge R&D programs which will have no less military value for their increase in general value.

Another administrative burden to the IR&D Program is the establishment of negotiated dollar ceilings for IR&D costs. It is recommended that such ceilings be eliminated and in their place the criterion of reasonableness should be applied by the government. Valid, real, competitive pressures exist for all companies bidding on defense contracts which act to prevent excessive expenditures for IR&D Programs. Companies spending too much will quickly find themselves losing defense contracts and will adapt their programs to represent a reasonable fraction of their defense revenue.

The elimination of potential military relationship and of negotiated ceilings in IR&D Programs will greatly strengthen these programs as a factor in the technical innovation process in the United States.

# APPENDIX A

## SUPPORTING ANALYSES FOR ADDITIONAL TAX CHANGES FOR POLICY CONSIDERATION

The following analyses support some of the additional tax policy changes that the Committee believes should be considered.

### CREDIT FOR INVESTMENT IN R & D FACILITIES[1]

A permanent increase in the investment tax credit from 10 to 20 percent for all business expenditures on buildings and equipment used in R & D activities would benefit innovation. Because taxpayers can reduce their tax liability only by increasing their investment spending, the investment tax credit is one of the most cost-effective tax incentives that government has at its disposal. Supplemental credits also have the advantage of involving no significant departures from existing tax arrangements.

The double tax credit should not be incremental; that is, it should not be limited to investment in excess of some base-period amount. To do so would reward businesses that have lagged in invention and innovation and penalize businesses that have been the most active innovators, especially businesses that maintain a high but relatively constant level of expenditures year after year. The incremental approach would also tend to produce cyclical and potentially destabilizing flows of research and development, the timing of which would depend on the availability of the credit as well as the need for the activity.

### FASTER DEPRECIATION FOR PATENTS

At present, the cost of acquiring another's patent must be capitalized and depreciated over the useful life of the patent. This creates two problems: First, unless the taxpayer can prove otherwise, there is a presumption that the seventeen-year legal life of a patent is also its economic or useful life. Yet, because of the pace of modern technological change, this is seldom the case; most patents become obsolete or widely imitated after a few years. Second, patents must be depreciated using the straight-line method and are not eligible for the accelerated depreciation (e.g., sum of the years' digits, declining balance) afforded other assets.

---

[1] In the U.S. tax code, the term *research and experimental* is used to define R & D activities.

Other intangible items of technology are treated even less favorably. Know-how, secret processes, and goodwill must be capitalized when acquired from third parties. Such items are not subject to depreciation because their useful lives are not readily ascertainable. This results in significant capital outlay without any available method of capital recovery during the period of use.

The overall result is that externally acquired patents and other intangible technology are treated less favorably than internally developed technological assets, for which research and experimental expenditures may be immediately expensed under Section 174. This discourages the purchase and commercialization of inventions and thereby limits the market opportunities of small inventors. Equal treatment would also help alleviate the waste of unutilized patents and intangible technology.

This Committee therefore suggests that externally acquired patents and intangible technology could both be subject to accelerated depreciation over a period of ten years or the demonstrated life, whichever is less. This would benefit successful inventors and innovators both indirectly by lowering one input cost and directly by facilitating the sale and commercialization of new technology.

## REQUIRE ONLY DIRECTLY RELATED R & D TO BE DEDUCTED FOR FOREIGN-SOURCE INCOME

For some time, U.S. tax law has required that a U.S. multinational corporation which incurs research and experimental costs in the United States must treat as a deduction from its foreign-source income the portion of such expenses that is properly allocated and apportioned to foreign-source income. Recently, this requirement was significantly extended by Treasury Regulation Section 1.861-8, which introduced certain new allocation and apportionment formulas that may require a percentage of research and development incurred by the U.S. parent company to be treated as a deduction from foreign-source income without showing any direct relationship to foreign activities or resulting financing benefit overseas.

The effect of this reallocation is to reduce the parent company's taxable foreign-source income (as computed for U.S. tax purposes) and increase its taxable U.S.-source income. As a result, a smaller foreign tax credit is available to the parent company to offset its U.S. tax liability on foreign-source income, and more tax is paid on U.S.-source income. Frequently, this means the effective loss of the deduction for research and development and increased worldwide taxation of U.S. multinational corporations. Furthermore, because foreign-based competitors of American firms operating abroad are not subject to such restrictions on the deductibility of their R & D expenditures, American multinationals are at a competitive disadvantage in foreign markets with a resultant decrease in their foreign earnings capabilities. The regulations thus diminish an American multinational's incentive to engage in research at home and concurrently diminish its earnings capabilities abroad at a time when both increased U.S. research and increased remittance of foreign earnings are in the national interest. U.S. multinationals may be able to deal with this dual problem by shifting some of their research effort to foreign facilities and thereby possibly obtaining a tax deduction from the foreign government, but such a solution works against nurturing more U.S. invention and innovation. Accordingly, the Committee urges that only the portion of a U.S. parent company's R & D expenses directly related and traceable to foreign earnings be treated as a deduction from foreign-source income.

## PROVIDE TAX CREDIT FOR SUPPORT OF UNIVERSITY RESEARCH

An imaginative approach to improving financing of basic research and fostering closer business-university relations would be provided by a tax credit for corporations that contribute to university research. There are several ways in which such a proposal could be formulated. The basic idea would be to make contributions for broad, nonproprietary research carried out by universities eligible for the credit. The mechanism would be similar to that of the investment tax credit; that is, a corporation could deduct a certain amount for each dollar of eligible contribution from its federal income tax liability.

The proposal has advantages: First, it would aid very basic research, the kind that is most difficult to finance. Second, it would add valuable diversity to sources of university finance. Third, it would help to make industry more conscious of basic research and its industrial applications. However, there are also disadvantages: It is not easy to estimate how much additional research funding would result. Second, it might be an unstable source of university finance, going up and down with corporate profit rates. On balance, however, we believe that this tax credit proposal merits serious study.

## ENCOURAGE INVENTIVE ACTIVITY IN SMALL BUSINESS

Although small businesses play an important role in invention and innovation, a variety of problems limit their activities in these fields. Capital-intensive research activity and commercial-scale innovation frequently require substantial production, financial, and marketing resources that are beyond the reach of small firms. Therefore, there is a need for public policy measures to assist small firms without removing them from the market forces.

Several provisions of the U.S. Internal Revenue Code currently favor small businesses and may indirectly provide incentives for inventive effort. Among those provisions are taxation of initial levels of corporate earnings below $100,000 at rates substantially below the basic tax rate; ordinary loss deductions up to $100,000 when small business corporation stock becomes worthless; avoidance of double taxation on the dividends paid to shareholders of regulated investment companies, particularly venture capital companies; and Subchapter S treatment[2] of the income from certain corporations with no more than fifteen shareholders.

In addition, the Committee believes the following three tax incentives that would encourage high-risk inventive efforts of small enterprises and would also encourage more business investment in general. These suggestions are discussed here in order of national priority.

## RELAXATION OF LIMITATIONS ON USE OF CAPITAL LOSSES TO OFFSET ORDINARY INCOME

Inventors would be more willing to undertake inventive and innovative risks if financial losses incurred as a result could be used to completely offset income earned elsewhere or in other time periods, with a resulting overall tax reduction.[3] Loss offsets currently permitted under the U.S. Internal Revenue Code are not complete. Corporate taxpayers may use capital losses only to offset capital gains and, at that, within a limited carry-over and carry-back period. Individual taxpayers may also deduct up to $3,000 of their net capital losses (all short-term losses plus 50 percent of long-term losses) from ordinary income each year, but according to a recent U.S. Treasury analysis, the limitation still results in significant undeducted capital losses in all income classes.[4]

The Committee believes that the deductibility of capital losses against ordinary income could be increased for all taxpayers. This would help to diminish the considerable risk incurred by investors in enterprises seeking radically new processes or products, a significant percentage of which end up as financial failures.

---

[2] Subchapter S treatment means that the income and losses of the corporation are passed directly through to corporate investors and are not taxed at the corporate level. This amounts to elimination of the double taxation of Subchapter S corporate income. At the same time, other advantages of incorporation, such as limited liability, are preserved.

---

[3] A thorough review of the theoretical and empirical analyses of investor behavior with respect to business risk is provided by Y. Ijiri in Chapter 3 of *Tax Policies for R & D and Technological Innovation* (Pittsburgh: Carnegie-Mellon University, 1976).

---

[4] National Science Board, *Science Indicators*, 1976 (Washington, D.C.: National Science Foundation, 1977).

## EXTENSION OF CARRY-OVER AND CARRY-BACK PERIOD

The government should also take steps to increase the ability of small businesses that frequently have little or no income as a result of heavy R & D expenditures to benefit from the investment tax credit, accelerated depreciation, and other tax measures designed to encourage investment by lowering or deferring taxes on current income. This could be accomplished by extending carry-back provisions for both net tax operating losses and investment tax credits from the current three years to seven years (the current maximum period for loss carry-overs).

## ENLARGEMENT OF ALLOWABLE SCOPE OF SUBCHAPTER S CORPORATIONS

Amendment of the provisions for Subchapter S corporations to allow as many as 100 investors instead of the present maximum of 15 would be a step toward improving the availability of venture capital. This would help new corporate enterprises to surmount the difficulties of raising sufficient front-end seed capital simply by increasing the number of sources they could tap.

# APPENDIX B

## SUPPORTING ANALYSES FOR ADDITIONAL SUGGESTED CHANGES IN PATENT POLICY

The analyses in this appendix supports the suggested changes in patent policy outlined in Chapter 6.

### SPECIAL PATENT APPEALS COURT

Honest irreconcilable differences occasionally dictate the need to litigate patent questions. Unfortunately, there are sometimes material differences in legal precedents about patents among different judicial circuits. The precedents in one circuit may indicate a favorable result for the plaintiff, but the precedents in another circuit may not. Because the result of the litigation can be affected or even determined by the circuit selected, the selection process, including the practice of the parties attempting to select the circuit that will be most favorable to their point of view, frequently becomes a major added complication.

**We recommend a single court of patent appeals. Such a court would help eliminate much of the costly procedural maneuvering at the early stages of patent litigation now devoted solely to selection of the forum.** A single court of appeals would also contribute to greater certainty in the predicted outcome of patent litigation. That is, a single court would tend to develop a more cohesive body of precedents than the many independent circuit courts do and might reduce the number of patent suits brought.

Although there might be difficulties in the handling of issues ancillary to patent validity and infringement, we believe that a single court would contribute far more to the value of the patent system than it would detract from it.

### COMPUTER SOFTWARE PROTECTION

An increasing proportion of money invested in research and development is being applied to the information processing field. The resulting computer programming (software) has not been afforded the same legal protection given to other technology. Although some software may find limited protection under the patent system or copyright, much appears to be protected by its owners as trade secrets.

There have been numerous recommendations that the present copyright statute be amended to protect computer programs. One such amendment is set forth by the National Bureau of Standards and the National Commission on New Technological Uses of Copyrighted Works, which made the following recommendation to the President of the United States on July 31, 1978:

> The new copyright law should be amended (1) to make it explicit that computer programs, to the extent that they embody an author's original creation, are proper subject matter of copyright; (2) to apply to all computer uses of copyrighted programs by the deletion of the present Section 117; and (3) to assure that rightful processors of copies of computer programs can use or adapt these copies for their use.

**We urge that the recommendation of the National Bureau of Standards and the National Commission on New Technological Uses of Copyrighted Works be implemented.**

## PROTECTION OF PROCESS PATENTS AGAINST IMPORTS

A problem of patent enforcement which may discourage investment in process development, is the powerlessness of a United States patent over products produced abroad by that process and brought into this country. A proceeding currently before the U.S. International Trade Commission may lead to the exclusion of such goods, but it has not proven satisfactory to many patent owners. Unlike the United States, other major countries do allow a process patent to be enforced through an infringement suit against foreign-made goods brought into the home country of the patent.

**In order to encourage investment in process innovation, we believe that patent owners should be able to enforce patents against goods made abroad by use of the patented processes and then imported into the United States.** Such a right would not bar the importation of identical goods not made by the patented process, but would secure to the patent owners what should be rightfully theirs.

## ADJUSTING TO PATENT TERM OF GOVERNMENT-CAUSED DELAYS

Government regulatory delays in clearing products for commercialization often shorten the effective life of a patent and undermine its value. This is particularly true in the cases of agricultural chemical products (which must be registered by the EPA) and pharmaceutical products (which must be cleared by the FDA). It is also true in other fields, for example, in the electronics industry in connection with the standardization of a particular color television system by the Federal Communications Commission.

The reduced values of such patents occur because, in many cases, the patent becomes effective a number of years before the products themselves can appear on the market. Thus, the patents expire relatively early in the commercial lives of the products, which enables potential infringers to get into the field quite soon without having to design around the patents or develop competing new products. Furthermore, the time in which the innovators of the products can recoup their investments is shortened.

**In our opinion patent owners should be able to receive extensions on their patent equal to any government regulatory delays.** To provide balance, there should be some maximum length of time for any extension as well as a precise definition of the kinds and causes of regulatory delays that might qualify patents for extensions. An alternate approach would be to allow patent owners to delay the effective date of their patent protection until any necessary government approvals for marketing the patented products are obtained.

## GOVERNMENT CONTRACTOR POLICY

Until the 1970s, much (if not most) of such federal funding was spent on military and space programs. In recent years, there has been a growing emphasis on development programs in other areas, such as energy. It is clearly in the national interest that the technology developed by federal funding be made available for use by private industry in providing products and services for the general public. In many areas of government, however, patent policies have been instituted either through statute or through agency rule making, which seem to discourage commercialization of the results of federally funded research and development.

Experience has shown that the company handling the government contract is most likely to carry the results of government-funded research and development to the marketplace. If the contractor must turn the rights back to the government or if competitors can quickly copy the product (e.g., by reverse engineering) without any patent deterrent, there is much less reason for the contractor to risk funds in commercialization. The same principle applies to the results of government-funded research and development done by nonprofit contractors such as universities. Unless the universities get substantial rights from patents, there is absolutely no incentive for them to establish technology transfer and patent programs that may lead to commercialization of the research.

To encourage the use of government-funded research and development for commercial products, contractors should, in most instances, receive title to the inventions and patents made under government contract. However, the government should be able to require the contractor to license others in certain circumstances (e.g., if the contractor fails to produce enough products to supply the market).

In some instances, it may be best for the government to take the title. For example, a regulatory agency may fund a development and then adopt it as a national standard for commercial products. To prevent economic dislocation in such a case, it is desirable that all competitors receive the same royalty-free right to use the required technology or feature.

Government patent policy could be further extended to improve procedures through which the government can more promptly and more readily recognize valid patents bearing on government activities. It is also important to discourage any contracting practices needlessly tending to appropriate privately developed rights of the contractor in existing patents or data.

## GOVERNMENT-OWNED PATENTS

A substantial portion of government R & D funding goes to support laboratories and other activities that are integral parts of government agencies. The unclassified technology developed by these laboratories has normally been patented, presumably to make sure that it becomes known to the public.

However, the administration of such patents presents a situation that is conceptually less straightforward than administration of patents generated from contractor-developed technology. Not having developed the technology, the commercially oriented engineers are likely to have, or at least to envision, different ways of accomplishing the same end. Thus, there is at least some question whether internally generated government patents serve as an effective tool for commercialization.

Such patents can certainly discourage manufacturers if they fear that the government will claim that they have the patent right to the product or process. **The best course is probably to license the internally generated government patents on a royalty-free basis to all domestic manufacturers.** In that way, the patents would serve to publish the technology and at the same time would not deter commercialization. (The patents could, however, serve to protect efficient domestic production from foreign dumping; and in the rare instance where exclusive licensing might be needed to elicit money for commercialization, such licensing could be contemplated.)

In many instances government patents are taken out solely for defensive reasons (i.e., to publish results of the technical work without any thought of the patents having commercial value). In view of the very large number of patent applications filed by the government each year on inventions made by government employees (averaging 1,332 applications a year for fiscal years 1963 to 1975), we believe that the work load of the PTO could be substantially reduced if this so-called defensive filing were eliminated by the government agencies.

**We suggest the establishment of a technical journal for the publication of selected government inventions. Such a journal would publish descriptions of those inventions on which patents are not needed for national security purposes.** In this way, the government would be protected against adverse claims by later inventors, the technology would be made available to interested parties, the government inventors would receive recognition, the PTO work load would be greatly reduced, and the government patenting costs would be decreased.

## OBJECTIVES OF THE COMMITTEE FOR ECONOMIC DEVELOPMENT

For thirty-five years, the Committee for Economic Development has been a respected influence on the formation of business and public policy. CED is devoted to these two objectives:

*To develop, through objective research and informed discussion, findings and recommendations for private and public policy which will contribute to preserving and strengthening our free society, achieving steady economic growth at high employment and reasonably stable prices, increasing productivity and living standards, providing greater and more equal opportunity for every citizen, and improving the quality of life for all.*

*To bring about increasing understanding by present and future leaders in business, government, and education and among concerned citizens of the importance of these objectives and the ways in which they can be achieved.*

CED's work is supported strictly by private voluntary contributions from business and industry, foundations, and individuals. It is independent, nonprofit, nonpartisan, and nonpolitical.

The two hundred trustees, who generally are presidents or board chairmen of corporations and presidents of universities, are chosen for their individual capacities rather than as representatives of any particular interests. By working with scholars, they unite business judgment and experience with scholarship in analyzing the issues and developing recommendations to resolve the economic problems that constantly arise in a dynamic and democratic society.

Through this business-academic partnership, CED endeavors to develop policy statements and other research materials that commend themselves as guides to public and business policy; for use as texts in college economics and political science courses and in management training courses; for consideration and discussion by newspaper and magazine editors, columnists, and commentators; and for distribution abroad to promote better understanding of the American economic system.

CED believes that by enabling businessmen to demonstrate constructively their concern for the general welfare, it is helping business to earn and maintain the national and community respect essential to the successful functioning of the free enterprise capitalist system.

# CED BOARD OF TRUSTEES

*Chairman*
FLETCHER L. BYROM, Chairman
Koppers Company, Inc.

*Vice Chairmen*
WILLIAM S. CASHEL, JR., Vice Chairman
American Telephone & Telegraph Company
E. B. FITZGERALD
Milwaukee, Wisconsin
PHILIP M. KLUTZNICK
Klutznick Investments
RALPH LAZARUS, Chairman
Federated Department Stores, Inc.
WILLIAM F. MAY, Chairman
American Can Company
ROCCO C. SICILIANO, Chairman
Ticor

*Treasurer*
CHARLES J. SCANLON, Vice President
General Motors Corporation

A. ROBERT ABBOUD, Chairman
The First National Bank of Chicago
RAY C. ADAM, Chairman and President
NL Industries, Inc.
WILLIAM M. AGEE, Chairman and President
The Bendix Corporation
ROBERT O. ANDERSON, Chairman
Atlantic Richfield Company
WILLIAM S. ANDERSON, Chairman
NCR Corporation
ROY L. ASH, Chairman
AM International, Inc.
ORIN E. ATKINS, Chairman
Ashland Oil, Inc.
THOMAS G. AYERS, Chairman
Commonwealth Edison Company
ROBERT H. B. BALDWIN, President
Morgan Stanley & Co. Incorporated
JOSEPH W. BARR, Corporate Director
Washington, D.C.
ROSS BARZELAY, President
General Foods Corporation
HARRY HOOD BASSETT, Chairman
Southeast Banking Corporation
ROBERT A. BECK, Chairman
Prudential Insurance Co. of America
WILLIAM O. BEERS, Chairman
Kraft, Inc.
GEORGE F. BENNETT, President
State Street Investment Corporation
JACK F. BENNETT, Senior Vice President
Exxon Corporation
JAMES F. BERE, Chairman
Borg-Warner Corporation
DAVID BERETTA, Chairman
Uniroyal, Inc.

JOHN C. BIERWIRTH, Chairman
Grumman Corporation
HOWARD W. BLAUVELT
Consultant and Director
Conoco Inc.
WILLIAM W. BOESCHENSTEIN, President
Owens-Corning Fiberglas Corporation
DEREK C. BOK, President
Harvard University
CHARLES P. BOWEN, JR., Honorary Chairman
Booz, Allen & Hamilton, Inc.
ALAN S. BOYD, President
AMTRAK
ANDREW F. BRIMMER, President
Brimmer & Company, Inc.
ALFRED BRITTAIN III, Chairman
Bankers Trust Company
THEODORE F. BROPHY, Chairman
General Telephone & Electronics Corporation
R. MANNING BROWN, JR., Chairman
New York Life Insurance Co., Inc.
JOHN L. BURNS, President
John L. Burns and Company
OWEN B. BUTLER, Vice Chairman
The Procter & Gamble Company
FLETCHER L. BYROM, Chairman
Koppers Company, Inc.
ALEXANDER CALDER, JR., Chairman
Union Camp Corporation
PHILIP CALDWELL, Vice Chairman and President
Ford Motor Company
ROBERT D. CAMPBELL, Chairman
Newsweek, Inc.
ROBERT J. CARLSON, Group Vice President
United Technologies Corporation
RAFAEL CARRION, JR., Chairman and President
Banco Popular de Puerto Rico
THOMAS S. CARROLL, President
Lever Brothers Company
FRANK T. CARY, Chairman
IBM Corporation
WILLIAM S. CASHEL, JR., Vice Chairman
American Telephone & Telegraph Company
FINN M. W. CASPERSEN, Chairman
Beneficial Corporation
JOHN B. CAVE, Senior Vice President,
  Finance and Administration
Schering-Plough Corporation
HUGH M. CHAPMAN, Chairman
Citizens & Southern National Bank of South Carolina
DAVID R. CLARE, President
Johnson & Johnson
WILLIAM T. COLEMAN, JR., Senior Partner
O'Melveny & Myers
*EMILIO G. COLLADO, President
Adela Investment Co., S.A.

*Life Trustee

ROBERT C. COSGROVE, Chairman
Green Giant Company

RICHARD M. CYERT, President
Carnegie-Mellon University

W. D. DANCE, Vice Chairman
General Electric Company

D. RONALD DANIEL, Managing Director
McKinsey & Company, Inc.

JOHN H. DANIELS, Chairman
National City Bancorporation

RONALD R. DAVENPORT, Chairman
Sheridan Broadcasting Corporation

RALPH P. DAVIDSON, Vice President
Time Inc.

R. HAL DEAN, Chairman and President
Ralston Purina Company

WILLIAM N. DERAMUS III, Chairman
Kansas City Southern Industries, Inc.

JOHN DIEBOLD, Chairman
The Diebold Group, Inc.

ROBERT R. DOCKSON, Chairman
California Federal Savings and Loan Association

EDWIN D. DODD, Chairman
Owens-Illinois, Inc.

VIRGINIA A. DWYER, Vice President and Treasurer
American Telephone & Telegraph Company

W. D. EBERLE, Special Partner
Robert A. Weaver, Jr. and Associates

WILLIAM S. EDGERLY, Chairman and President
State Street Bank and Trust Company

ROBERT F. ERBURU, President
The Times Mirror Company

THOMAS J. EYERMAN, Partner
Skidmore, Owings & Merrill

WALTER A. FALLON, Chairman
Eastman Kodak Company

FRANCIS E. FERGUSON, President
Northwestern Mutual Life Insurance Company

JOHN T. FEY, Chairman
The Equitable Life Assurance Society
of the United States

JOHN H. FILER, Chairman
Aetna Life and Casualty Company

WILLIAM S. FISHMAN, Chairman
ARA Services, Inc.

E. B. FITZGERALD
Milwaukee, Wisconsin

JOSEPH B. FLAVIN, Chairman
The Singer Company

CHARLES F. FOGARTY, Chairman
Texasgulf Inc.

CHARLES W.L. FOREMAN, Senior Vice President
United Parcel Service

DAVID L. FRANCIS, Chairman
Princess Coals, Inc.

*WILLIAM H. FRANKLIN
Chairman of the Board (Retired)
Caterpillar Tractor Co.

DON C. FRISBEE, Chairman
Pacific Power & Light Company

RICHARD M. FURLAUD, Chairman
Squibb Corporation

DONALD E. GARRETSON
Vice President, Finance
3M Company

CLIFTON C. GARVIN, JR., Chairman
Exxon Corporation

RICHARD L. GELB, Chairman
Bristol-Myers Company

LINCOLN GORDON, Senior Research Fellow
Resources for the Future, Inc.

THOMAS C. GRAHAM, President
Jones & Laughlin Steel Corporation

HARRY J. GRAY, Chairman and President
United Technologies Corporation

JOHN D. GRAY, Chairman
Hart Schaffner & Marx

JOHN D. GRAY, Chairman
Omark Industries, Inc.

WILLIAM C. GREENOUGH
Trustee and Chairman, CREF Finance Committee
TIAA-CREF

DAVID L. GROVE, President
U.S. Council of the I.C.C.

JOHN D. HARPER, Retired Chairman
Aluminum Company of America

SHEARON HARRIS, Chairman
Carolina Power & Light Company

FRED L. HARTLEY, Chairman and President
Union Oil Company of California

ROBERT S. HATFIELD, Chairman
Continental Group, Inc.

GABRIEL HAUGE
Director and Retired Chairman
Manufacturers Hanover Trust Company

BARBARA B. HAUPTFUHRER, Corporate Director
Huntingdon Valley, Pennsylvania

PHILIP M. HAWLEY, President
Carter Hawley Hale Stores, Inc.

H. J. HAYNES, Chairman
Standard Oil Company of California

H. J. HEINZ II, Chairman
H. J. Heinz Company

LAWRENCE HICKEY, Chairman
Stein Roe & Farnham

RODERICK M. HILLS
Latham, Watkins and Hills

WAYNE M. HOFFMAN, Chairman
Tiger International, Inc.

ROBERT C. HOLLAND, President
Committee for Economic Development

LEON C. HOLT, JR., Vice Chairman
Air Products and Chemicals, Inc.

FREDERICK G. JAICKS, Chairman
Inland Steel Company

EDWARD R. KANE, President
E. I. du Pont de Nemours & Company

HARRY J. KANE, Executive Vice President
Georgia-Pacific Corporation

DONALD P. KELLY, President
Esmark, Inc.

J. C. KENEFICK, President
Union Pacific Railroad Company

*Life Trustee

JAMES L. KETELSEN, Chairman
Tenneco, Inc.

TOM KILLEFER, Chairman
United States Trust Company of New York

E. ROBERT KINNEY, Chairman
General Mills, Inc.

PHILIP M. KLUTZNICK
Klutznick Investments

RALPH LAZARUS, Chairman
Federated Department Stores, Inc.

FLOYD W. LEWIS, President
Middle South Utilities, Inc.

FRANKLIN A. LINDSAY, Chairman
Itek Corporation

JOHN A. LOVE, President
Ideal Basic Industries

J. PAUL LYET, Chairman
Sperry Corporation

RICHARD W. LYMAN, President
Stanford University

RAY W. MACDONALD, Honorary Chairman
Burroughs Corporation

BRUCE K. MacLAURY, President
The Brookings Institution

MALCOLM MacNAUGHTON, Chairman
Castle & Cooke, Inc.

JOHN MACOMBER, President
Celanese Corporation

G. BARRON MALLORY
Jacobs Persinger & Parker

WILLIAM A. MARQUARD, Chairman and President
American Standard Inc.

AUGUSTINE R. MARUSI
Chairman, Executive Committee
Borden Inc.

WILLIAM F. MAY, Chairman
American Can Company

JEAN MAYER, President
Tufts University

*THOMAS B. McCABE
Chairman, Finance Committee
Scott Paper Company

C. PETER McCOLOUGH, Chairman
Xerox Corporation

GEORGE C. McGHEE
Corporate Director and
   former U.S. Ambassador
Washington, D.C.

JAMES W. McKEE, JR., Chairman
CPC International Inc.

CHAMPNEY A. McNAIR, Vice Chairman
Trust Company of Georgia

E. L. McNEELY, Chairman
The Wickes Corporation

RENE C. McPHERSON, Chairman
Dana Corporation

J. W. McSWINEY, Chairman
The Mead Corporation

BILL O. MEAD, Chairman
Campbell Taggart Inc.

CHAUNCEY J. MEDBERRY III, Chairman
Bank of America N.T. & S.A.

RUBEN F. METTLER, Chairman
TRW, Inc.

CHARLES A. MEYER
Senior Vice President, Public Affairs
Sears, Roebuck and Co.

FREDERICK W. MIELKE, JR., Chairman
Pacific Gas and Electric Company

LEE L. MORGAN, Chairman
Caterpillar Tractor Co.

STEVEN MULLER, President
The Johns Hopkins University

ROBERT R. NATHAN, Chairman
Robert R. Nathan Associates, Inc.

EDWARD N. NEY, Chairman
Young & Rubicam Inc.

WILLIAM S. OGDEN, Executive Vice President
The Chase Manhattan Bank

NORMA PACE, Senior Vice President
American Paper Institute

THOMAS O. PAINE, President
Northrop Corporation

EDWARD L. PALMER
Chairman, Executive Committee
Citibank, N.A.

RUSSELL E. PALMER, Managing Partner
Touche Ross & Co.

VICTOR H. PALMIERI, Chairman
Victor Palmieri and Company Incorporated

DANIEL PARKER
Washington, D.C.

JOHN H. PERKINS, President
Continental Illinois National Bank
   and Trust Company of Chicago

HOWARD C. PETERSEN
Philadelphia, Pennsylvania

C. WREDE PETERSMEYER
Bronxville, New York

MARTHA E. PETERSON, President
Beloit College

PETER G. PETERSON, Chairman
Lehman Brothers Kuhn Loeb, Inc.

JOHN G. PHILLIPS, Chairman
The Louisiana Land and Exploration Company

CHARLES J. PILLIOD, JR., Chairman
The Goodyear Tire & Rubber Company

JOHN B. M. PLACE, President
Crocker National Bank

DONALD C. PLATTEN, Chairman
Chemical Bank

EDMUND T. PRATT, JR., Chairman
Pfizer Inc.

LEWIS T. PRESTON, President
Morgan Guaranty Trust Co. of New York

JOHN R. PURCELL, Executive Vice President
CBS, Inc.

R. STEWART RAUCH, JR.
Chairman, Executive Committee
General Accident Group of Insurance Companies

DONALD T. REGAN, Chairman
Merrill Lynch & Co., Inc.

DAVID P. REYNOLDS, Chairman
Reynolds Metals Company

*Life Trustee

JAMES Q. RIORDAN, Executive Vice President
Mobil Oil Corporation

BRUCE M. ROCKWELL, Chairman
Colorado National Bank

FELIX G. ROHATYN, General Partner
Lazard Freres & Company

WILLIAM M. ROTH
San Francisco, California

JOHN SAGAN, Vice President-Treasurer
Ford Motor Company

TERRY SANFORD, President
Duke University

CHARLES J. SCANLON, Vice President
General Motors Corporation

HENRY B. SCHACHT, Chairman
Cummins Engine Company, Inc.

ROBERT M. SCHAEBERLE, Chairman
Nabisco Inc.

J. L. SCOTT, Chairman
Great Atlantic & Pacific Tea Company

D. C. SEARLE, Chairman
  Executive Committee
G. D. Searle & Co.

ROBERT V. SELLERS, Chairman
Cities Service Company

ROBERT B. SEMPLE, Chairman
BASF Wyandotte Corporation

MARK SHEPHERD, JR., Chairman
Texas Instruments Incorporated

RICHARD R. SHINN, President
Metropolitan Life Insurance Company

ROCCO C. SICILIANO, Chairman
Ticor

ANDREW C. SIGLER, Chairman and President
Champion International Corporation

WILLIAM P. SIMMONS, President
First National Bank & Trust Company

L. EDWIN SMART, Chairman
Trans World Airlines

DONALD B. SMILEY, Chairman
R. H. Macy & Co., Inc.

RICHARD M. SMITH, Vice Chairman
Bethlehem Steel Corporation

ROGER B. SMITH, Executive Vice President
General Motors Corporation

ELVIS J. STAHR
Senior Counselor and Past President
National Audubon Society

CHARLES B. STAUFFACHER, President
Field Enterprises, Inc.

EDGAR B. STERN, JR., President
Royal Street Corporation

J. PAUL STICHT, Chairman
R. J. Reynolds Industries, Inc.

GEORGE A. STINSON, Chairman
National Steel Corporation

*WILLIAM C. STOLK
Weston, Connecticut

WILLIS A. STRAUSS, Chairman
Northern Natural Gas Company

DAVID S. TAPPAN, JR., Vice Chairman
Fluor Corporation

WALTER N. THAYER, President
Whitney Communications Corporation

W. BRUCE THOMAS, Executive Vice President
  Accounting and Finance
United States Steel Corporation

WAYNE E. THOMPSON, Senior Vice President
Dayton Hudson Corporation

HOWARD S. TURNER
Chairman, Executive Committee
Turner Construction Company

L. S. TURNER, JR., Executive Vice President
Texas Utilities Company

THOMAS A. VANDERSLICE, President
General Telephone & Electronics Corporation

ALVIN W. VOGTLE, JR., President
The Southern Company, Inc.

SIDNEY J. WEINBERG, JR., Partner
Goldman, Sachs & Co.

GEORGE WEISSMAN, Chairman
Philip Morris Incorporated

WILLIAM H. WENDEL, President
Kennecott Copper Corporation

GEORGE L. WILCOX, Director-Officer
Westinghouse Electric Corporation

*FRAZAR B. WILDE, Chairman Emeritus
Connecticut General Life Insurance Company

J. KELLEY WILLIAMS, President
First Mississippi Corporation

*W. WALTER WILLIAMS
Seattle, Washington

MARGARET S. WILSON, Chairman
Scarbroughs

RICHARD D. WOOD, Chairman and President
Eli Lilly and Company

*Life Trustee

# HONORARY TRUSTEES

E. SHERMAN ADAMS
New Preston, Connecticut
CARL E. ALLEN
North Muskegon, Michigan
JAMES L. ALLEN, Hon. Chairman
Booz, Allen & Hamilton, Inc.
FRANK ALTSCHUL
New York, New York
O. KELLEY ANDERSON
Boston, Massachusetts
SANFORD S. ATWOOD
Lake Toxaway, North Carolina
JERVIS J. BABB
Wilmette, Illinois
S. CLARK BEISE, President (Retired)
Bank of America N.T. & S.A.
HAROLD H. BENNETT
Salt Lake City, Utah
WALTER R. BIMSON, Chairman Emeritus
Valley National Bank
JOSEPH L. BLOCK, Former Chairman
Inland Steel Company
ROGER M. BLOUGH
Hawley, Pennsylvania
FRED J. BORCH
New Canaan, Connecticut
MARVIN BOWER, Director
McKinsey & Company, Inc.
THOMAS D. CABOT, Hon. Chairman of the Board
Cabot Corporation
EDWARD W. CARTER, Chairman
Carter Hawley Hale Stores, Inc.
EVERETT N. CASE
Van Hornesville, New York
HUNG WO CHING, Chairman
Aloha Airlines, Inc.
WALKER L. CISLER
Detroit, Michigan
STEWART S. CORT, Director
Bethlehem Steel Corporation
GARDNER COWLES, Hon. Chairman of the Board
Cowles Communications, Inc.
GEORGE S. CRAFT
Atlanta, Georgia
JOHN P. CUNNINGHAM
Hon. Chairman of the Board
Cunningham & Walsh, Inc.
ARCHIE K. DAVIS, Chairman (Retired)
Wachovia Bank and Trust Company, N.A.
DONALD C. DAYTON, Director
Dayton Hudson Corporation
DOUGLAS DILLON, Chairman, Executive Committee
Dillon, Read and Co. Inc.
ALFRED W. EAMES, JR., Director
Del Monte Corporation
ROBERT W. ELSASSER
New Orleans, Louisiana
EDMUND FITZGERALD
Milwaukee, Wisconsin

WILLIAM C. FOSTER
Washington, D.C.
JOHN M. FOX, Retired Chairman
H.P. Hood, Inc.
CLARENCE FRANCIS
New York, New York
GAYLORD FREEMAN
Chicago, Illinois
PAUL S. GEROT, Hon. Chairman of the Board
The Pillsbury Company
CARL J. GILBERT
Dover, Massachusetts
KATHARINE GRAHAM, Chairman
The Washington Post Company
WALTER A. HASS, JR., Chairman
Levi Strauss and Co.
MICHAEL L. HAIDER
New York, New York
TERRANCE HANOLD
Minneapolis, Minnesota
J. V. HERD, Director
The Continental Insurance Companies
WILLIAM A. HEWITT, Chairman
Deere & Company
OVETA CULP HOBBY, Chairman
The Houston Post
GEORGE F. JAMES
Cos Cob, Connecticut
HENRY R. JOHNSTON
Ponte Vedra Beach, Florida
GILBERT E. JONES, Retired Vice Chairman
IBM Corporation
THOMAS ROY JONES
Consultant, Schlumberger Limited
FREDERICK R. KAPPEL
Sarasota, Florida
CHARLES KELLER, JR.
New Orleans, Louisiana
DAVID M. KENNEDY
Salt Lake City, Utah
JAMES R. KENNEDY
Essex Fells, New Jersey
CHARLES N. KIMBALL, Retired President
Midwest Research Institute
HARRY W. KNIGHT, Chairman
Hillsboro Associates, Inc.
SIGURD S. LARMON
New York, New York
DAVID E. LILIENTHAL, Chairman
Development and Resources Corporation
ELMER L. LINDSETH
Shaker Heights, Ohio
JAMES A. LINEN, Consultant
Time Inc.
GEORGE H. LOVE
Pittsburgh, Pennsylvania
ROBERT A. LOVETT, Partner
Brown Brothers Harriman & Co.

ROY G. LUCKS
Del Monte Corporation
FRANKLIN J. LUNDING
Wilmette, Illinois
IAN MacGREGOR, Honorary Chairman
 and Chairman, Executive Committee
AMAX Inc.
FRANK L. MAGEE
Stahlstown, Pennsylvania
STANLEY MARCUS, Consultant
Carter Hawley Hale Stores, Inc.
JOSEPH A. MARTINO, Hon. Chairman
N L Industries, Inc.
OSCAR G. MAYER, Retired Chairman
Oscar Mayer & Co.
L. F. McCOLLUM
Houston, Texas
JOHN A. McCONE
Pebble Beach, California
JOHN F. MERRIAM
San Francisco, California
LORIMER D. MILTON
Citizens Trust Company
DON G. MITCHELL
Summit, New Jersey
ALFRED C. NEAL
Harrison, New York
J. WILSON NEWMAN
Chairman, Finance Committee
Dun & Bradstreet Companies, Inc.
AKSEL NIELSEN
Chairman, Finance Committee
Ladd Petroleum Corporation
JAMES F. OATES, JR.
Chicago, Illinois
W. A. PATTERSON
Honorary Chairman
United Air Lines
EDWIN W. PAULEY, Chairman
Pauley Petroleum, Inc.
MORRIS B. PENDLETON
Vernon, California
JOHN A. PERKINS
Berkeley, California
RUDOLPH A. PETERSON
President (Retired)
Bank of America N.T. & S.A.
PHILIP D. REED
New York, New York
MELVIN J. ROBERTS
Denver, Colorado
AXEL G. ROSIN, Retired Chairman
Book-of-the-Month Club, Inc.
GEORGE RUSSELL
Bloomfield Hills, Michigan

E. C. SAMMONS
Chairman of the Board (Emeritus)
The United States National Bank of Oregon
JOHN A. SCHNEIDER
New York, New York
ELLERY SEDGWICK, JR.
Cleveland Heights, Ohio
LEON SHIMKIN, Chairman
Simon and Schuster, Inc.
NEIL D. SKINNER
Indianapolis, Indiana
ELLIS D. SLATER
Landrum, South Carolina
S. ABBOT SMITH
Boston, Massachusetts
DAVIDSON SOMMERS
Vice Chairman
Overseas Development Council
ROBERT C. SPRAGUE
Hon. Chairman of the Board
Sprague Electric Company
FRANK STANTON
New York, New York
SYDNEY STEIN, JR., Partner
Stein Roe & Farnham
ALEXANDER L. STOTT
Fairfield, Connecticut
FRANK L. SULZBERGER
Chicago, Illinois
CHARLES P. TAFT
Cincinnati, Ohio
C. A. TATUM, JR., Chairman
Texas Utilities Company
ALAN H. TEMPLE
New York, New York
CHARLES C. TILLINGHAST, JR.
Managing Director, Capital Markets Group
Merrill Lynch Pierce Fenner & Smith, Inc.
LESLIE H. WARNER
Darien, Connecticut
ROBERT C. WEAVER
Department of Urban Affairs
Hunter College
JAMES E. WEBB
Washington, D.C.
J. HUBER WETENHALL
New York, New York
A. L. WILLIAMS
Ocean Ridge, Florida
*WALTER W. WILSON
Rye, New York
ARTHUR M. WOOD, Director
Sears, Roebuck and Co.
THEODORE O. YNTEMA
Department of Economics
Oakland University

*CED Treasurer Emeritus

## TRUSTEES ON LEAVE FOR GOVERNMENT SERVICE

W. GRAHAM CLAYTOR, JR.
Deputy Secretary of Defense
A. L. McDONALD, JR.
Staff Director
The White House

JOHN C. SAWHILL
Deputy Secretary
Department of Energy

# RESEARCH ADVISORY BOARD

*Chairman*
THOMAS C. SCHELLING
John Fitzgerald Kennedy School of Government
Harvard University

GARDNER ACKLEY
Henry Carter Adams University
Professor of Political Economy
Department of Economics
The University of Michigan

MARTIN FELDSTEIN, President
National Bureau of Economic Research, Inc.

CARL KAYSEN
Vice Chairman and Director of Research
The Sloan Commission on Government
 and Higher Education

ANNE O. KRUEGER
Professor of Economics
The University of Minnesota

ROGER G. NOLL, Chairman and Professor of Economics
Division of the Humanities and Social Sciences
California Institute of Technology

ANTHONY G. OETTINGER, Director
Program on Information Resources Policy
Harvard University

GEORGE L. PERRY, Senior Fellow
The Brookings Institution

WILLIAM POOLE, Professor of Economics
Brown University

ARNOLD R. WEBER, Provost
Carnegie-Mellon University


# CED PROFESSIONAL AND ADMINISTRATIVE STAFF

ROBERT C. HOLLAND
President

SOL HURWITZ
Senior Vice President

KENNETH McLENNAN
Vice President and Director
 of Industrial Studies

CLAUDIA P. FEUREY
Director of Information

FRANK W. SCHIFF
Vice President
 and Chief Economist

KENNETH M. DUBERSTEIN
Vice President,
 Director of Business-
 Government Relations
 and Secretary, Research
 and Policy Committee

S. CHARLES BLEICH
Vice President, Finance
 and Secretary, Board of Trustees

R. SCOTT FOSLER
Vice President and
 Director of Government Studies

ELIZABETH J. LUCIER
Comptroller

*Research*
SEONG H. PARK
Economist

*Business-Government Relations*
MARGARET J. HERRE
Staff Associate

*Conferences*
RUTH MUNSON
Manager

*Information and Publications*
HECTOR GUENTHER
Assistant Director

SANDRA KESSLER
Assistant Director

CLAUDIA MONTELIONE
Publications Manager

*Finance*
PATRICIA M. O'CONNELL
Deputy Director

HUGH D. STIER, JR.
Associate Director

*Administrative Assistants
to the President*
THEODORA BOSKOVIC
SHIRLEY R. SHERMAN

# STATEMENTS ON NATIONAL POLICY
# ISSUED BY THE RESEARCH AND POLICY COMMITTEE

## PUBLICATIONS IN PRINT

Stimulating Technological Progress *(January 1980)*

Helping Insure Our Energy Future:
 A Program for Developing Synthetic Fuel Plants Now *(July 1979)*

Redefining Government's Role in the Market System *(July 1979)*

Improving Management of the Public Work Force:
 The Challenge to State and Local Government *(November 1978)*

Jobs for the Hard-to-Employ:
 New Directions for a Public-Private Partnership *(January 1978)*

An Approach to Federal Urban Policy *(December 1977)*

Key Elements of a National Energy Strategy *(June 1977)*

The Economy in 1977-78: Strategy for an Enduring Expansion *(December 1976)*

Nuclear Energy and National Security *(September 1976)*

Fighting Inflation and Promoting Growth *(August 1976)*

Improving Productivity in State and Local Government *(March 1976)*

*International Economic Consequences of High-Priced Energy *(September 1975)*

Broadcasting and Cable Television:
 Policies for Diversity and Change *(April 1975)*

Achieving Energy Independence *(December 1974)*

A New U.S. Farm Policy for Changing World Food Needs *(October 1974)*

Congressional Decision Making for National Security *(September 1974)*

*Toward a New International Economic System:
 A Joint Japanese-American View *(June 1974)*

More Effective Programs for a Cleaner Environment *(April 1974)*

The Management and Financing of Colleges *(October 1973)*

Strengthening the World Monetary System *(July 1973)*

Financing the Nation's Housing Needs *(April 1973)*

Building a National Health-Care System *(April 1973)*

*A New Trade Policy Toward Communist Countries *(September 1972)*

High Employment Without Inflation:
 A Positive Program for Economic Stabilization *(July 1972)*

Reducing Crime and Assuring Justice *(June 1972)*
Military Manpower and National Security *(February 1972)*
The United States and the European Community *(November 1971)*
Improving Federal Program Performance *(September 1971)*
Social Responsibilities of Business Corporations *(June 1971)*
Education for the Urban Disadvantaged:
   From Preschool to Employment *(March 1971)*
Further Weapons Against Inflation *(November 1970)*
Making Congress More Effective *(September 1970)*
Training and Jobs for the Urban Poor *(July 1970)*
Improving the Public Welfare System *(April 1970)*
Reshaping Government in Metropolitan Areas *(February 1970)*
Economic Growth in the United States *(October 1969)*
Assisting Development in Low-Income Countries *(September 1969)*
\*Nontariff Distortions of Trade *(September 1969)*
Fiscal and Monetary Policies for Steady Economic Growth *(January 1969)*
Financing a Better Election System *(December 1968)*
Innovation in Education: New Directions for the American School *(July 1968)*
Modernizing State Government *(July 1967)*
\*Trade Policy Toward Low-Income Countries *(June 1967)*
How Low Income Countries Can Advance Their Own Growth *(September 1966)*
Modernizing Local Government *(July 1966)*
Budgeting for National Objectives *(January 1966)*
Educating Tomorrow's Managers *(October 1964)*
Improving Executive Management in Federal Government *(July 1964)*

---

\*Statements issued in association with CED counterpart organizations in foreign countries.

## CED COUNTERPART ORGANIZATIONS IN FOREIGN COUNTRIES

Close relations exist between the Committee for Economic Development and independent, nonpolitical research organizations in other countries. Such counterpart groups are composed of business executives and scholars and have objectives similar to those of CED, which they pursue by similarly objective methods. CED cooperates with these organizations on research and study projects of common interest to the various countries concerned. This program has resulted in a number of joint policy statements involving such international matters as energy, East-West trade, assistance to the developing countries, and the reduction of nontariff barriers to trade.

| | |
|---|---|
| **CE** | Círculo de Empresarios<br>*Serrano Jover 5-2°, Madrid 8, Spain* |
| **CEDA** | Committee for Economic Development of Australia<br>*139 Macquarie Street, Sydney 2001,*<br>*New South Wales, Australia* |
| **CEPES** | Europäische Vereinigung für<br>Wirtschaftliche und Soziale Entwicklung<br>*Reuterweg 14, 6000 Frankfurt/Main, West Germany* |
| **IDEP** | Institut de l'Entreprise<br>*6, rue Clément-Marot, 75008 Paris, France* |
| 経済同友会 | Keizai Doyukai<br>(Japan Committee for Economic Development)<br>*Japan Industrial Club Bldg.*<br>*1 Marunouchi, Chiyoda-ku, Tokyo, Japan* |
| **PSI** | Policy Studies Institute<br>*1-2 Castle Lane, London SW1E 6DR, England* |
| **SNS** | Studieförbundet Näringsliv och Samhälle<br>*Sköldungagatan 2, 11427 Stockholm, Sweden* |